SOLAR SUSPICIONS

ISBN: ISBN: 9798864118375

Imprint: Independently published

Table of Contents

Preface

Technosignature - noun [C] US/ˈtek.noʊˌsɪg.nə.tʃɚ/ UK/ˈtek.nəʊˌsɪg.nə.tʃər/

Something that shows the use of technology in a place, which is a sign of the past or present existence of intelligent living things there:

- Most searches for technosignatures have focused on radio signals.
- NASA began supporting the scientific search for technosignatures as part of the larger search for extraterrestrial life.

Cambridge Dictionary (Cambridge University Press & Assessment 2023) - TECHNOSIGNATURE definition | Cambridge English Dictionary

Technosignature – Exoplanets or biosignatures identified through perturbations of atmospheric composition or other planetary qualities, in observable ways.

NASA Astrobiology Strategy 2015

Technosignatures - The search for structured electromagnetic signals as well as a wide variety of other evidence of intentional manipulation of matter and energy—from alien megastructures to industrial pollution, or nighttime lighting systems on distant worlds.

Scientific American - Caleb A. Scharf, July 2, 2021

Astrobiology is a young field of science with plenty of room for interpretation and imagination. Spurred on by movies and television, we are only a few generations into this mind-expanding science as the data pours in from concerned citizens to top Astrophysicists that we may not be alone in this universe. The hunt for Technosignatures is now openly discussed in all quarters of society under the terms of UFOs, UAPs, SFOs, and many more acronyms now common world-wide. This book is focused on Technosignatures and is the reason I gave three definitions above as the term is new within this field of study.

I hope you choose this book to read because you are interested in the latest research in astrobiology or space research in general. You may also want to know if there is definitive proof life exists beyond our planet (this book will help you answer that question). Perhaps you are the kind who enjoys reading about creation versus evolution theories. Maybe you feel the scientific community (the dark side - beyond top secret) is withholding information from the public for our own safety. Others of you may be interested in how religions have not answered the basic question if other life exists in the universe all the while building and buying observatories. This book answers that too. You may end up burning this book after reading it because it offends your faith in God or some of you may create a new religion around the premise I propose and miss the main point entirely. Either way, what you are about to read will stretch your imagination, challenge your faith, force you to consider a new paradigm of life that may end up being a bit uncomfortable from prior teachings.

There is a ying and yang to living that is displayed in all aspects of this journey we call life. I ask you to keep an open mind and equally weigh both sides of the arguments offered in the coming chapters. For those of you who love politics, this book may not be for you as lately the right and left side seem to believe one side can live without the other. The best way to grow in knowledge is by listening to both sides of an argument and weighing them without bias or tilting the scales if possible. Usually, with enough credible information presented, one argument will make more sense than the other. It is how we weigh new facts or listen to arguments to determine if there is enough merit to accept the new piece of information as a new fact or worthy of more consideration.

It helps to know from what point of view a person writes or conveys an idea. The perspective of this author (me) gives this story some flavor and a bit of my own bias. It is up to you to weigh what is presented as makes sense or completely foreign and nonsensical. To set the stage properly I will attempt to introduce myself with enough depth and background to leave no doubt what shaped my life. Looking back over my life I can see clearly why I needed to write this book given the journey I am on. This book is a culminating point for me to convey some of my extraordinary findings that will add to the astrobiology community in a meaningful way. I have been on an eight-year road of exploration that never disappoints and continues to power my imagination of the life around us.

I invite you to pick and choose the chapters to read since some folks don't have time to waste on things that may not matter to them, so I titled my chapters to be as accurate of a description as possible. If you chose to read this book straight through, the advantage will be understanding my viewpoints based on experiences and education with better clarity as I reveal what has laser focused my attention for the last eight years (since 2015) and expanded my understanding of the amount of Extraterrestrial life around us.

To answer the basic question up front first; "Are we alone in the universe? I can answer that with an unequivocal "No".

For those who are time-management challenged, this answer should suffice if you do not have time to continue reading to find out how I came to this conclusion. Thank you for choosing this book and if you learn or see something new, perhaps it will spur you on your own journey.

Introduction

Allow me to introduce myself. I was born on 27 June 1961 and grew up in Phoenix Arizona in a low to mid-income Christian home. My parents did a fantastic job raising my older brother and I, now that I look back on all of it. Everything I experienced as a child and a teenager is now culminating in how I arrived at my latest discovery of life beyond this planet.

After high school and with a couple of college semester hours completed, I decided it was time to get married at the ripe old age of 19 (fortunately, she said "Yes"). The economy was in shambles in 1981 but passion won out over wisdom, and I married a beautiful childhood friend that I re-met in college. Only a year later and we were expecting our first child. It was a poor time to start a family as it turns out, so I started looking toward military service to support my family. At least they supported my desire to go to school and provide health care that I did not have previously. Arriving at the storefront called Armed Forces Recruiting Station, the Air Force recruiters welcomed me but projected a wait time to ship out at a year and a half. That was way too long for me as I needed a better paying job immediately. So, I slipped over next door to the Army guys and asked the fateful question "Can I join and ship out soon?". SSG (Staff Sergeant) Black gladly shook my hand and brought me right in. He was kind and to the point, "What do you want to do?". I simply said, "I guess anything that is challenging and transferable to the civilian market later". The list of jobs I was given repeated between "Cook" or "Truck Driver". With a large grin, SSG Black promised a ship-out time of two weeks if I would just sign on the dotted line. Instantly, 4 years of service seemed like a lifetime, so I stalled him by asking if there were other jobs that were more complex in nature. That conversation was extended for over two months as his frustration with me started to show. SSG Black finally offered some rare electronics job in Military Intelligence field that required a top-secret clearance 33S MOS (Military Operation Skill). I accepted it immediately, however, he had to find someone to explain the details as he had little documentation to go with the listing. I was taken into a special room for another briefing on the job when a new guy in uniform came in and sat way too close to me, like smell-your-breath too close. He began to describe the job specifics, but then added I should be fully aware if my background check turned up anything disqualifying, I'd have to accept any job the Army offered me for the remaining four years this position required. Before I could answer he began listing all the deadly sins found in the Bible, and a few new ones our twisted culture has come up with that sounded just as bad as the Ten Commandments. I told him I had done nothing in my past related to anything he listed, and said I was a good candidate to cross over into Mormonism since I didn't drink coffee either. He didn't laugh or change facial expressions but simply stared into my soul and declared they will find whatever it is I may be hiding. Without blinking throughout the entire conversation, I was betting he was reptilian, but I couldn't prove it. I couldn't be happier when he left the room but knew I had found the right job; 33S – Electronic Warfare Intercept Systems Tactical Repairer. The MOS was focused on repairing anything within the Army system.

It turns out I still had to wait 3 weeks before shipping out 5 February 1984, but I had a job that sounded worth learning. Basic Training came and went at Ft. Dix NJ, but Advanced Individual Training was a year-long at Fort Devens, MA. It also took that long to complete the required background investigation to obtain the top-secret clearance needed to do my work. The school seemed longer than it was and taught every electronic theory and repair technique from coffee pots to top secret surveillance/communications equipment whose capabilities are still covered in secrecy today. Again, I never intended to stay for the past four years but even though I made it to the end of my contract with Uncle Sam, I found the cost of living even worse than when I went in. President Regan, my Commander in Chief in 1988 was still trying to recover our crippled economy as I decided to re-enlist for a stint of 6 years just to feed my family. However, this time I had a better plan and that was to go into Officer Candidate School (OCS). It is the Army's way of commissioning officers other than West Point or ROTC. The clear advantage is the prior enlisted time I had acquired in the 5 and a half years before my accepted into OCS. This meant I had earned a respect level from the troops more readily than from another commissioning source as a new Second Lieutenant. I am sure I just pissed-off a few ring-knockers reading this book, but they know deep down it's true.

The odds of being accepted into OCS and then Initial Entry Rotary Wing school (helicopter school) afterwards were in the low, single digits. Those odds only made me want it more, so I applied to see where life would take me. The answer to that question would take several more books that I don't intend to write. After completing OCS, I was assigned to flight training at Ft. Rucker, AL. They used UH-1 helicopters at the time that all seem to have a Vietnam story to go with each of the 90 airframes on base. Even the old civilian flight instructors would go into a "There I was" story mid-flight. Only a couple of times did our Vietnam veteran flight instructors began to sweat profusely during their in-flight flashbacks that would hasten your search for an emergency landing area just in case. However, there were no better trainers and training that went on at Ft. Rucker regardless of the size and depth of the egos found on one Army base. Even the flight students seemed to find time to spin a fictitious "There I was" stories to some willing listener in a bikini on the shores of Panama City Beach Florida on base at the Officers Club.

After flight school ended, I was assigned to Ft. Hood TX to 2/1 Cavalry, and then 1/6th CAV, 6th Cavalry Brigade. The US just began the push into Iraq during Operation Desert Storm 2 August 1990 when I took my first evaluation flight around the pattern at Hood Army Airfield. I spent four years at Ft. Hood and never left for combat or transferred to another unit. Instead, after my four years were up, I got out of the service all together in hopes of saving a failing marriage of 13 years and resettled back home in Phoenix. The marriage didn't survive but another year, and I spun my wheels for a few years or so to regain my sense of self that included a year-long rebound marriage that also met an early demise.

Nothing in life, especially after two failed marriages, leaves one with enough confidence to think my personal life was on track. My Dad and Stepmom were gracious enough to pick up the pieces after marital strike two. After all, I had attended Bible college, been ordained as a Southern Baptist deacon at the age of 24 and enjoyed a recent successful aviation career and a BS degree to boot. I was more convinced than ever the reason I liked Dr. Spock on the original Star Trek series was because I identified with his character better than the others. However, Dr. Spock was unmarried for a very long time (simply logical).

I tried to get back into the military as soon as possible, and especially after my marital misfortunes. I knew I could rebuild a career from where I came and retire close to my original planned timeframe when I started flying. Finding an Active Guard Reserve position in the Arizona Army National Guard became my plan "A" since it would count as active time towards my retirement which now only required another 10 years to complete. As it turned out, I did find an open position and continued my aviation officer career as one of the many distinguished AZ National Guard soldiers I was privileged to work with. I also married for the third time to a wonderful lady named Lory who had the legal and business background to rival the best entrepreneurs in Arizona. Her skill set would come in handy throughout our lives for everyone who knew her. Her two kids and my two kids were living on their own by the time we married, but like most households we were home base for a couple of them returning for a place to stay temporarily.

I was given a company command position of an attack helicopter unit consisting of six Apache AH-64A's and three Kiowa OH-58's Scout helicopters. I had never been surrounded by a more professional group of soldiers as they showed a mountain of patience while I learned the ropes of company command. Learning to fly the Apache was one of my toughest assignments as it was the most complicated system of the three airframes I was qualified to fly. The main challenge was the split visual system of one monocle in my right eye of the camera point of view off the nose of the aircraft, and the uncovered left eye scanning the interior dashboard of flight system information. At night the contrast between the two was startling at first and challenging to say the least. What helped me learn the system was a simple motivation of survival. If you make a mistake from the cockpit the cascading effect can be swift and unforgiving. After my command and later promotion to the rank of Major, I entered the phase of my career in supporting command operations and logistics as a "Staffer". This meant the next command opportunity would be years down the road, if ever, since I only had a few years left before retirement.

Near the end of my Army career, I asked to leave the Guard and join an active-duty unit in Iraq during Operation Iraqi Freedom in 2004. I wanted to retire knowing I had applied my skills in support of a combat mission I had trained for all my Army life. The other motivation was my son was already over there fighting in a Cavalry unit near Baghdad in Forward Operating Base Falcon. I was granted my wish, and a Captain friend of mine, Pat O'Toole joined me in our one and only combat tour with the 1st CAV Division in the 4th Aviation Brigade already in place at Camp Taji, Iraq in the Summer of 2004. Pat flew a lot of missions in and out of Iraq, but since I was a Major, the Brigade command thought I'd be better suited to work with their Logistics (S4) section. I spent most of my combat time making sure everyone else fighting had what they needed to be successful in their missions. No flight time for me during my combat tour but I did use the skill set they needed most in ensuring the resupply was on time and at the right amounts.

I reached the end of my combat tour along with the 1st CAV and headed back home after a much-needed break of being in harm's way 24/7. I rejoined my national guard unit in 2005 and retired in 2006 as an Army Aviation Major. Looking back at the 10- year commitment the Air Force was asking for to attend their academy I had to laugh at the irony of my late commitment of 22 years to the Army during my retirement ceremony. Later in 2006 I picked up a job overseas as a contractor working for Kellogg, Brown & Root in Camp Speicher, Iraq. I was hired as a Logistics Coordinator but later worked up the ladder to a Liaison Officer working directly for the Regional Program Manager. Our region covered 9 camps in the Northern part of Iraq. I worked as the go between for KBR and the military to ensure real-time support issues were addressed, modified, improved, or scrapped to mirror the needs of the military operations at the time.

I ended up staying overseas 6 more years in either Iraq or Afghanistan with 5 different companies through 5 different contracts. My last contract was with Engility Corporation in 2011-2012 flying 22-foot, helium filled Aerostats with an extremely capable MX15i camera system hanging just below at 1500 feet above our camp. Our job was to observe the surrounding local town to understand life cycles and normal activities or patterns of the population. What this allowed us to do was compare normal activities to unusual or threatening activities against the base by way of individuals or groups firing weapons at the base or driving car bombs into the base main gate. The whole multi-million-dollar surveillance system was centered on the capability of the camera operator (civilian contractor such as I was) to have studied the local population, understand what normal patterns of life were and compared those patterns to individuals or groups who didn't fit the pattern or were outwardly threatening the local population or our forward operating base. This work contributed to my skill set that I will cover later on in this book.

Many times, I felt like looking for that proverbial "needle in the haystack" as our team scanned 24/7 the movements and commerce happening within a few feet of our walls to 15 miles away. Our personal motivation was unspoken as everyone knew our failure to detect threats in a timely manner meant we ran the risk of harm to ourselves and others. The risk of death became a constant motivation to avoid it by being hyper situationally aware. Only by operating at the maximum level of our own capabilities and continually sharpening our skills to find needles in haystacks did we offer enough time and distance from incoming attacks to counter them before they adversely effected our base operations. It turns out this was my second job where learning imagery analysis in real time meant a higher rate of survival. My first imagery analysis job was flying an Apache attack helicopter and now a static blimp system scanning for real-time threats to increase your odds for survival.

One of the guys I worked with in Afghanistan on the Aerostat team got a job offer to fly drones with another company at a higher pay rate than what we already enjoyed. He had only 50 hours of flight time in Cessna 152 with a recent private pilot's license, but they accepted him just the same. I thought, I have over one thousand hours of flight time in three different turbine helicopter airframes, so I would easily qualify for a similar job. In early 2013 I left Afghanistan to be interviewed for a drone job in Yuma Arizona thinking if my buddy with low flight time could do it, I could do the same. As it turns out, I could not. I failed to account for my buddy's ability to fly RC (remote control) aircraft when he was a kid whereas I lacked RC transmitter time using just your thumbs. Sure enough, during the hands-on portion of the interview they handed me a toy transmitter box with two sticks and asked me to fly a quarter-million-dollar drone currently flying 2000 feet overhead. I did my best to get a grip on what the sticks controlled on the drone but translating my years of manned flight control through my thumbs failed miserably. In fact, I think I made the interviewer nervous as he was sweating more than I during my attempt at controlled flights, which never happened. I simply didn't have the chops to make the cut. I drove away from Yuma knowing I was not going to get a second interview and returned to Phoenix not knowing what my future would hold. After walking in the house to announce the odds of getting that job, my wife took the information in stride and simply said "Why not learn to fly drones"? I laughed a bit trying to understand what she meant. She reiterated, "Go to school to learn whatever it is that you need to learn". After we did some research, we found only three schools at the time offering Unmanned Aircraft System bachelor's degrees; Kansas State, North Dakota State, and Embry-Riddle Aeronautical University in Daytona beach Florida. We had family not 20 miles from Daytona Beach so the decision to move was an easy one. We moved to Florida in August 2013 for me to get schooled up in a new and blossoming sUAS (small Unmanned Aircraft Systems) industry.

Education

I had picked up a Bachelor of Science degree in liberal studies from Excelsior College while I was still in the Regular Army back in 1993. It was the only way the Army would keep me as an officer as I had only accumulated a pile of semester hours during my civilian time and a few more during my early enlisted days. However, I never applied them towards a degree. The new Army rules a year prior mandated I get an accredited four-year degree or leave the service. I was given 12 months to pull this off, however, I already had 134 semester hours sitting around waiting to be applied to a 124 semester-hour degree. By the time the dust settled just 6 months later I had a brand-new BS degree right before the cut-off time to stay in the service. Ironically, I retired early from the Regular Army into the Army Inactive Reserve the next year trying to save a marriage my wife and I had slowly destroyed. In 1994 we left to live back home in Phoenix for the next year until I put a stake into the marriage and divorced the next year after leaving the service. That was in 1995. In 1996 I remarried and divorced in 1998 after a typical rebound scenario. It wasn't until 2000 that I found a lady who understood me better than myself when looking at the whole person. Her name is Lory, and we married in 2001. This year (2023) we celebrated 22 years together. The next major timeline event for me was when I retired from military service in 2006 and went into overseas contracting until 2013.

In 2013 I had tried but failed to use my BS in Liberal Studies to acquire a dream job while working the last 6 and ½ years overseas as a contractor. That was a long stretch of time for being away from home, so I was planning to do one last contract by interviewing for a drone flying job. I thought the drone/UAS interview in Yuma would be a cakewalk, it was not. My wife suggested I turn my interview failure into a guiding star to learn the skills I lacked to fly these drones. After offering my negative impression to her suggestion, her subtle brilliance finally penetrated my resistive stance, and we began to research four-year universities offering a BS in drones. It didn't take long to find one as there were a total of three universities in the country at the time offering a four-year degree. So, Lory and I drove out to Embry-Riddle Aeronautical University in Daytona Beach Florida to try this new plan of getting an updated education in an industry that was merely science-fiction when I received my first BS degree 20 years earlier. The move was uneventful as we seemed to find our groove early on as to the routine on-campus life demanded. We found a home in a beautiful part of Florida called Deland in a Norman Rockwell neighborhood that was truly breath-taking. Our house faced a 10-acre lake the whole neighborhood was developed around. The word pristine does not describe adequately the care taken by the HOA for our housing development and the view we enjoyed from our front porch overlooking the lake was unbelievable.

The lake across the street from our house played an ominous role later in my UAS education by devouring my first DJI Vision 2+ drone that cost me $1,200. It only took 3.5 seconds to fall 40 feet helplessly into our beautiful neighborhood lake after a cracked blade guard split enough to stop a blade from spinning. To those who know what blade guards are, never fly a drone with a broken blade guard unless you can handle the emotional pain that awaits you. I was not emotionally ready and got a little misty-eyed walking back into the house with a useless transmitter in one hand and a clinched fist in the other. My loving wife asked a simple question upon my entrance into our home that I had previously tried to answer in my mind during my 39 steps back to the front door- "Where's your drone"? In a monotoned voice and without breaking my stride to the bedroom I said, "In the lake". With the wisdom of Solomon, she did not say another word after offering up an "I am sorry". She knew I needed time to process this series of events I created through a Swiss Cheese model of human error as authored by James Reason. Had I only had the chance to hang out with this "James Reason" guy a few days before my UAS drowned in the lake, I now call "Lake Sorrow", things could have been different. This was my first aviation accident but since it involved a drone, it only hurt my pride and pocketbook as I learned this valuable lesson: always pay attention to the details no matter how small because your loss could be significant.

My classes at Embry-Riddle were challenging for more reasons than one. First, I was old. Second, I was older than most of my professors. Third, my age seemed to reveal a lack of understanding of this thing called computer programming. The kids in my classes (I say kids because I had grandchildren almost their age), all seemed to have grown up with some intuitive programming knowledge. I, on the other hand, remember the punch cards and a room full of computers that would eat them to operate properly. To say computer programing was a weak area of my academic prowess would be a major under-statement. I used my student loans to find tutors who would help me understand programming and help me approach parity with my peers. That never happened but I did pass all my computer classes just the same. Embry-Riddle AU already held the reputation of being the world's elite pilot training institution and was quickly bringing the UAS training up to that same level.

Religion

Attending church since I was two weeks old, growing up learning what faith was and how it was defined, helped me understand science better. Yep, I said science. It takes faith to accept theories offered by schoolteachers or professors in a classroom setting where some theories lack empirical evidence. In a classroom you just sit and listen to someone persuade you they are speaking the truth. You may even read a book that supports the same concepts. Is it possible both religion and science could be leaning to one side or the other of this razor thin edge called truth? Absolutely! In fact, both groups of evolutionists and creationists point to the other and use the same criticism against each other in using proof comingled with theory/faith. Both groups use some faith to stitch theories into belief. It's why some folks call science the new religion and some call religion, well, religion.

I bring religion into this conversation as it has shaped my life as much as science. They both have similar mechanisms to gain one's trust, and in the end, when no one has solid answers to basic or complex questions, a step of faith is needed to envision the outcome. The main problem is science explains our current world and religion attempts to help us navigate this world as a bridge to the world to come. Everything connected to the world to come is predicated on how we manage our behavior, make decisions, and what we believe in our current world. Religion and science both agree we all die. However, science offers nothing beyond that point. Religion offers something for people that connects us to the next life. Since anything beyond the grave is based on belief or faith, however, when neither is present, one is left with only the here and now. This is where science excels as it offers proofs that turn theory into "facts" and builds on prior research to advance new theories that require more proofs. In religion this process starts with a belief (written or verbal) in something and then looks to validate the belief through the course of our lives. Religious proofs come in the form of archeology, astronomy, natural time clocks, personal experiences, and prophecy. Science offers prophecy as well but does so in the form of statistical predictions through linear regression models, Bayesian Theory, or statistical modeling. The prophecies of science end where death begins. For those who are believers, death ends where faith begins. This book only has one focus, and that is to prove both science and religion are both wrong in concealing the magnitude of life within our universe and creating the depth of societal ignorance and misinformation that we are the only ones that exist in the universe. God and science are much bigger than that.

Observatories by their nature are instruments for the exploration of our solar system and universe. Secular and religious institutions around the world use observatories to push our knowledge about our past and try to discover our place in the universe with new discoveries every year. Many religious institutions today run observatories on their campuses to include Protestant schools like Hasting College and Wheaton College as a part of their science curriculum. Likewise, private Christian schools such as Butler University offer degrees in astronomy by operating sophisticated telescopes on their respective campuses. Scientific exploration today is as common as breathing but not so in our recent past. The Roman Catholic church was notorious for persecuting anyone who didn't accept their science of an Earth centered universe such as the case against Galileo in the 17th century. The real story behind the 17th century history of church repression concerning astronomical studies was more of a question of questioning the church. The books by Nicholas Copernicus and Galileo Galilei were banned by the Catholic church as soon as they were published. Although both men were giants in their field, the church was not in the position to be questioned. It was the arrogance of the church, not the hatred of science that got these men in trouble. The church, more specifically the Roman Catholic religion, was the predominately held faith of the time and openly explored the skies and encouraged understanding of astronomy. However, the focus of the 17th century church during this period was more concerned with calendar accuracy to determine the precise date to celebrate Easter. Any other discovery not related to this gap in their knowledge was scorned as heretical since all other astronomical matters were already decided and blessed by the Pope. It would be many centuries later science would honor Copernicus and Galileo for their work as scientific reasoning gained momentum centuries later.

Paying Attention

After a year and a half as an ERAU student, I opened a weather web site one day to a looped weather pattern over Daytona Beach on 4 January 2015 only to see a large micro-burst like cloud development repeat in the loop however the entire state was enjoying a cloudless day. A microburst is usually a few miles in diameter and only lasts a few minutes, however this concentric circle was 50 to 60 miles across when it finally dissipated after an hour or two. I expanded the area on the map to see more of Florida and found another similar sized micro-burst developing about 50 miles south at the exact same time and both appeared and disappeared within the few hours of the loop. I expanded the map to see all of Florida and counted 4 to 5 concentric circles of massive micro-bursts across the whole state. All of the microbursts started and ended at the same time yet 50 to 100 miles from each other. I have a lot of weather experience as a pilot, and a deep respect for such destructive weather phenomenon as micro-bursts but to see one appear out of a clear day is unusual. To see 4 to 5, appear across an entire state at the exact same time is completely unbelievable, except I have a video record of the event. I showed a few pilot friends of mine at the time, and they all thought it was cool since no one had seen it before and then began reliving the weekend festivities they just endured. When they are not flying pilots can exhibit traits of social ADD if the conversation drifts away from a perfectly good "There I was" aviation story. I get the short-lived interest since the information I just shared with them does not present as a threat or something cooler than a good flying story.

Not dissuaded by their inability to link a cause to the phenomenon, I continue to pursue an investigative line of thought determined to find a reasonable source for the strange weather episode. Thinking more globally, I realized the size of the phenomenon may have a space component to it or at least be associated with a solar flare. I searched Space Weather on the internet and found a site called Space Weather Prediction Center published by the National Oceanic and Atmospheric Administration (NOAA). On the main page are sources of imagery used to determine what the sun is doing on any given day. One image caught my attention as a Blue-filtered Sun was presented on the screen. I later found out it was an image produced by the COR1 sensor on the Solar Helio-spheric (SOHO) satellite using Ultra-Violet frequencies at the 171 Angstrom wavelength. That said, Ultra-Violet frequencies can be colored post-production to identify the frequency band used to produce the image. NASA uses a color convention to delineate one sensor frequency from another for quick post processing reference. The Blue sun I was looking at on the main page was produced by what they call a COR-1 sensor (COR= Coronal). The SOHO also has a COR-2 sensor that is slightly off-set frequency-wise but is Red in color. Both frequencies choices are looking at the sun but detecting different material generated by the sun specific to that frequency. It sounds confusing because we were all taught the sun is simply a ball of Hydrogen burning through a fusion process with different densities as you move to the core. Not completely true. More on that subject to follow.

When I clicked on the blue sun it turned into a mini-looped movie. It showed 80 still pictures in rapid succession based on some predetermined cadence of time. This sensor displayed a wide field of view around the sun but blocks the sun directly with a mask showing only the streams of light or coronal ejections from the surface of the sun outward. What caught my attention this time was the sparkles of light in all shapes and sizes adding a festive look to the movie. It replicated the look and randomness of a sparkler lit in the background. Only certain light spots remained and turned out to be stars and planets in the shared scene. However, in one of the 80 frames that continued to flash before my eyes, an unusual pattern reappeared in just one frame that seemed out of place. I played with the movie controller to slow and stop-frame on the shape and it revealed a perfect Right-angle formed with two equal length lines. I thought the odds of two streaks seemingly connected to each other were remote but possible. If they were connected, I had a long road ahead of me to determine what they were and would involve asking a lot of smart people a few dumb questions. I was up for that since at my age embarrassment had lost its power over me when it comes to learning something new. I really thought I was seeing alien spacecraft, but hesitated realizing millions of other folks see these images for a living, and no one is sounding the alarm we are being invaded. I was pretty sure there was a good explanation waiting for me within the hollowed walls of academia.

My Space Psychology professor, Dr. Kring, was an absolute space enthusiast with little care to what others thought of his Star Wars and Star Trek toy collection in his office. His true love was all things "space" so I decided I would run my images from the SOHO by him to get his reaction. I had collected even more images with weird objects in the few weeks after my first discovery and caught Dr. Kring in his office for a quick Q & A session. I am sure he expected a class assignment question but was equally intrigued to see some strange objects within the satellite pics I ran past him. One of the objects looked like the silhouette of the USS Enterprise from Star Trek so we had some fun talking about that one. I asked if he could tell me what I was seeing in the images, but he admitted he was not sure. He then grinned and asked, "What do you think they are?" I knew what he was getting at, I stated it looked like something non-natural and perhaps intelligent. He smiled and said, "I can't tell, but the guys on the third floor should be able to help you." The third floor housed the classes and offices of the astronomy department. It was a logical next step.

Heading out to classes on campus January 20th, 2015, I decided to get on campus early and run to the astronomy department and ask my handful of dumb questions to kick-start this new line of education I was being drawn into. I had a few weird, shaped objects in 10 or so images and needed an expert to illuminate my mind on the subject. Speaking with the department receptionist I asked if there was anyone on the floor who knew anything about the satellite SOHO? Immediately, a gentleman peered around the corner of a hallway wall behind the receptionist and said, "I do!". It turned out to be Dr. John Hughes, associate professor of engineering physics and well versed in all things space. It was a slow day, and he had a few minutes to offer. We discussed the objects in the images as he one-by-one could identify the aberration in each except for three of the images. He looked intently at them again and said, "It's hard to say what these are". He asked, "What are you thinking they could be?" I composed myself to be reserved and calm as possible and said, "Maybe something more than natural, possibly intelligent". Without batting an eye, he encouraged me to keep exploring and get back to him if I found anything else worthy of discussion.

In the meantime, and over the next several months I spent more time researching the SOHO archives online downloading anything that looked weird. I started scanning the archives of all the sensors used by the SOHO, and even found a couple of other solar satellite systems, the SDO and STEREO-A&B satellites and started reviewing their archived images as well. Pretty soon I had to start organizing the growing amount of data I was pulling from these three systems. I set up separate files for each on my computer and decided to use their own date/time group architecture to name the images I was downloading. This way any picture I had in my file replicated the same name found in the NASA archives. I still use the same naming convention when processing images today as helps to point to others who want proof back to the source or location from the NASA archives from where they originated. Date-Time groups as names also makes creating data sets easy for future analysis.

Then one night I was scanning the NASA archives in my now usual manner and clicking through the COR2 sensor images in the SOHO website when I see this unusual spot on the far-Right side of the image. Like usual, I zoom in on the object only to find my first piece of evidence Extraterrestrial Vehicles could be real. An object containing multiple bright pixels in an intelligent shape was staring me in the face. My world changed. I felt my soul twist up and then released into a heap of unanswerable questions. Few things in life reach the depth of my soul as this did. Becoming a Christian at the young age of eight years old, getting married, having children all changed the trajectory of my life and affected me at the deepest levels of my being. However, this event was emotionally significant because it was the proof I was looking for. I stopped scanning for objects for a few days to get my mind wrapped around what I found. I didn't know who to report this too as it was obvious a space vehicle cruising past a solar satellite located 1 million miles away from Earth. That is three times the distance away from Earth than the moon. Since it could not be space trash this object was clearly out of place as the oval-shaped pixel dots were screaming intelligent design. I kept this image to myself as a secret Holy grail until I could find more like it. The image below (page 20) is what I came across after a year of searching.

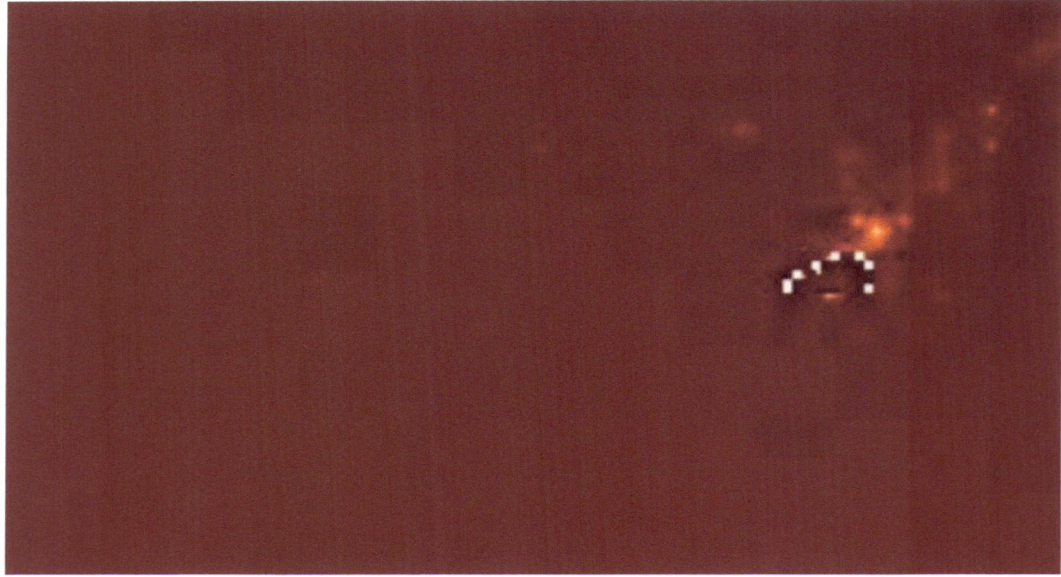

Courtesy of NASA SOHO Image Archive
(https://sohowww.nascom.nasa.gov/data/data.html). Image date/time/sensor: 19970201 0553 C2

Validating the Unknown

During lunch time in the main campus cafeteria, I would sit with my usual set of friends and every now and then show them a weird object in a NASA image just to see their reactions. Of course, I was immediately dubbed the old man with the Tinfoil hat. No matter, as some of the pics did what I hoped in making them look twice or start a discussion of "What ifs". One of my lunch friends, David Brock, recommended I talk to a professor, Dr. William Barott, a professor of electrical engineering and involved with the SETI project before arriving to teach at Embry-Riddle. SETI stands for Search for Extraterrestrial Intelligence. His office was in the Engineering building on campus and an easy walk from the Student Center. I made an appointment to introduce myself and show him a few images with hard to explain objects within various sensor frequencies from the SOHO, STEREO-A and SDO satellites. He looked through the images and began to discuss the possibilities of finding intelligent life through satellite sensors but felt he had not seen enough clear evidence within the data set I presented to convince him of such a discovery. However, he encouraged me to continue searching as he felt eventually, I could come across a piece of data that would be worthy of serious consideration.

My run through the academic gantlet seemed to leave me mildly encouraged while quietly disappointed no one jumped up and crowned me top Astrophysicist of the year with a Nobel Prize for disclosing one of man's greatest universal mysteries. Instead, I found myself more determined than before to find something, anything that would capture these guys' attention as much as this subject had captured mine. As I finished my Master of Science in Human Factors at Embry-Riddle AU I realized my path in life was deliberate, but I felt it was by someone else's design. I had no prior knowledge needed in this scientific discipline to approach this discovery with systematic methodology while using the highest levels of academic protocols to prove the slightest piece of evidence to be considered.

The reason there is a major divide between the those in academia and those who believe life exists beyond our planet is proof. However, a scientist without proof is naturally a skeptic because they have not been abducted. As it turns out, one alien abduction can change a whole lot of science on a personal level. The problem is truly one of the common experiences. How do you convince someone else the private experience you just had was real? Provide pictures or a video? Use other witnesses to the same event or experience? It takes more than word of mouth or a good written story. Even a good video of something unexplainable simply means you have something unexplainable. Because most of our scientific institutions are government funded, like NASA, whose focus is all things space, we would expect some attempt to explain the unexplainable in their own archived videos or images. Instead, there is dead silence. I get the quiet lapse of time to conduct further research before hanging their Doctoral degrees on something as crazy as Extraterrestrial life may exist in our universe, galaxy, solar system or on our planet. On the one hand, you would hope they, NASA, would at least acknowledge the possibility of intelligent life (greater than ours), could exist and why the evidence they do have may or may not fit that possibility. When the evidence is so overwhelming, I expect NASA to at least admit they have no other explanation for the event or image(s), therefore it could be an ET event or ET object.

In class, the professors would sometimes give a complex problem to solve as a group assignment and it would be interesting to hear the possible solutions offered in response. Many times, the professor would stop and say "Consider Occam's Razor" to focus our thoughts on the simplest solution that had the highest probability of being the best answer. However, when an unusual aerial craft is witnessed or video is being considered by the public, no official in the scientific community will offer ETs as the best possible solution. The resistance to utter the words "Extra Terrestrial" seems like a United States problem in contrast to other countries who handle the phenomenon more openly. I grew up in the aviation community flying Army helicopters for 17 years and found it unusual there was no reporting mechanism to report UFO sightings within the Federal Aviation Administration (FAA) or the military. In fact, the aviation culture in general was simply ignoring it. Most pilots had no interest in talking UFOs and would snicker a bit if the subject was brought up. Those conversations would later reveal a new nickname for you if you were the one who brought up the subject or were crazy enough to state out loud you believed. In my circle of aviation acquaintances, I don't recall anyone admitting to believing in UFOs,,,at least publicly. The closest anyone would get to this subject was by linking a story of a sighting to some distant friend or wacky relative. I wasn't going to find validation or much support in the aviation community even though I am sure there were plenty of closet believers.

Scientific Methods

I was blessed to go through four semesters of statistics during my Master of Science in Human Factors program and enjoyed each and every class. You were correct if you detected massive sarcasm in the sentence above. Even though statistics and data analysis are fundamentally the best way to dissect a problem, the mere reference to them still inculcates raised levels of anxiety in my soul. Just thinking about the effort required for me to grasp some of the fundamentals in my studies causes uncontrolled twitching in my right eye. Statistics did not come naturally to me, but I knew it was the Holy Grail proving something existed that officially didn't exist. Unlike chemistry or engineering that is founded on quantitative analysis using mathematics, imagery analysis takes a different road of qualitative analysis and then converts into quantitative analysis. The qualitative analysis is more subjective (i.e., "The photograph has a weird looking object in it") while quantitative analysis is more objective in nature (i.e., "I now have three photographs with weird looking objects in it").

Handling data analysis and statistics is like handling nitroglycerin while running a race to prove something new. If you mis-handle statistics in front of your professors, bad things happen. Like Jedi-Masters, Professors can detect cracks in your model(s) or numeric conclusion(s) and leave you intellectually bleeding in front of your classmates. Being called out as a statistical heretic can inflict unrecoverable damage to your reputation if you are not diligent enough to ensure numerical accuracy in your proofs. This is why few rookie statisticians venture out into the murky waters of this numerical science as there seems to be no lack of sharp shooters waiting in the shadows to publicly humiliate you into intellectual obscuration. The possibility of being accurate enough to avoid friendly fire from government sponsored scientists or those in academia is near impossible. Compound this risk by trying to prove alien life forms exist with statistics and you end up on an academic suicide mission. So, that is what I set out to do, survive the challenge of statistical proof of aliens in an academic environment.

If you recall the previous chapter titled "Education", then you will also recall most of my recent education came from the premier aeronautical institutions on the planet, Embry-Riddle Aeronautical University. It was in my Master of Science classes for Human Factors where I learned the power of statistics. In 2016 I had already collected several thousand images of what I thought were weird objects within NASA solar satellite photographs. I did notice one such satellite (STEREO –

A) offered a series of similar unusual objects that reappeared in the C-2 sensor images consistently. In the middle of my second statistics class in this major I decided to use my NASA satellite data set for my term paper. We had to choose a data set to run through a new data program just taught in the same class called "R" Studio. This program is so large and powerful major health corporations use it to validate medical studies. We were fortunate in that our professor, Dr. James Novack, was so well versed in this massive program he was able to teach us the program and the multi-variate statistical concepts at the same time. To say the least I was a bit apprehensive to walk into Dr. Novack's office to declare my data set for the class paper. However, instead of giggling at what I was attempting to prove with my data set, he simply guided my processes to ensure I ended up with an accurate result regardless of the outcome. It was a simple process to find if any intelligible wave forms resulted within a Histogram as my Null Hypothesis. A Null Hypothesis is an expected outcome whereas the Alternative Hypothesis is an opposite outcome.

My Statistics class was nearing its end, and my professor was curious about the result I had found in my data set of weird objects near the solar satellite. I shared the graph that resulted in a strange wave shape in the Histogram with Dr. Novack. I looked for an explanation or interpretation from Dr. Novack, but he simply stated there was enough anomalous shape to the Histogram that he recommended further research and deeper statistical analysis. This meant another semester of statistics would see another run at my data set, but the next level of analysis resulted in more than anyone expected.

Scientific Reason

Dr. Frank Drake was born on May 28th, 1930, in Chicago IL., and became a well-known radio astronomer and astrophysicist. He is currently involved in the search for extraterrestrial intelligence (SETI) at 89 years of age and offered up a theoretical equation in the early '60s known as Drake's Equation. It listed out a string of independent variables like the rate of galaxy creation, star creation, planet creation, life formation, intelligence formation, etc., that added up to a numerical possibility for life elsewhere in our own galaxy. Dr. Drake is quoted in many peer-reviewed articles in astrobiology and radio astronomy to include his famous equation for predicting Extraterrestrial life.

Drake's Equation is $N = R^* \bullet fp \bullet ne \bullet fl \bullet fi \bullet fc \bullet L$ and is used more as a discussion point than to determine an accurate number of intelligent societies elsewhere in our galaxy. The "N" part of the equation is the dependent variable or answer to the equation once all the independent variables or values are determined. The variables listed are the rate stars form that are suitable for sustaining intelligent life (R^*), the fraction of those same stars with planetary systems (fp), the number of those planetary systems that could sustain life (ne), the number of planets in those systems where life exists (fl), the planets containing life with emerging intelligence (fi), the number of intelligent societies capable of signal emanations into space (fc), and finally the length of time those civilizations exist (L). It is a fun way to logically extrapolate what the possibilities are we are not alone.

Fermi's Paradox takes the opposite point of view and offers up a simple counter to Drake's equation. It simply states that since no evidence exists of other life forms in the galaxy, then no other life forms exist. It is based on a Bayesian principle of priori knowledge being the foundation for a predictable outcome. Since no evidence currently is offered as proof of Extraterrestrial life, then the obvious prediction is none will be found in the future. However, this approach to science is defeatist as it encourages no further exploration to counter any theory and goes against the very nature of science; to explore. If everyone abided by Fermi's approach to science, there would be no man on the moon or satellites sent out into the edges of our solar system because it had not been done before.

The part of Dr. Drake's equation that caught my eye was the fact he constrained the equation to focus on Exoplanets alone. Exoplanets are simply those planets found in other solar systems, and specifically ones that could potentially sustain life. We already have space travel in space craft and live in space stations with only 100 years of history of flight on our planet. If we can accelerate our technology at that same speed for another 100 years, our means of travel would be as unimaginable to us as it was for our pioneers of flight thanks to the Wright brothers. By extending that same logic forward for another technological society who has advanced 1,000 years ahead of us, what kind of vehicles would they be traveling in? Would they be traveling in vehicles at all? In fact, a measure of a societal form of transportation may directly correlate with the advancement of their technology. If two different species of life, from two different locations in the galaxy, appeared on Earth at the same time with two different levels of technological advancements, they will most likely arrive by two different methods. One may materialize while the other pulls up in a slick looking spacecraft. Who knows, but the chances they would be equal in types of transportation is extremely low. The same applies to cloaking technology as we can cloak materials today, and for the last 10 years openly published these advancements for the world to see. Once again, applying the rate of advancement of this technology, it would be hard to determine just how far it will go in 100 years. For this reason, I determined Drake's equation was incomplete. It simply lacked the biggest variable of them all: **V**ehicles in **s**pace. So, I added it to the original equation:

$$N = R^* \bullet fp \bullet ne \bullet fl \bullet fi \bullet fc \bullet L \bullet \text{Vs}$$

This new equation now allows the astrobiologist or space enthusiast to include the entirety of space as a viable search domain. This theory states: If our stellar neighbors are 100 to 1,000 years more advanced, then they would already be space faring societies and possibly flying by just waiting for us to get our act together as a species.

Regarding cloaking, one can cloak materials all day long but there is a limit to what types of material can be cloaked. This limit exists until it doesn't because it is based on advances in our current technology. However, one hurdle that may never be resolved is the interaction of the material within its surrounding environment. If I drop a rock that is cloaked into the pond, it will still make the water ripple. It's how the Invisible Man gets defeated in the cartoons by simply changing the immediate environment of his surroundings to smoke or flour. Another hurdle is trying to cloak the process of atoms or molecules changing energy states. A good example would be cloaking a rocket ship before takeoff, but during takeoff cloaking the massive plume of energy escaping the backside of the thrusting system would be impossible. You would still see the flames and smoke associated with a rocket launch. Likewise, if the cloaked rocket ship flew through the chromosphere of the sun, you could see the effect of the plasma reshaping around the ship as it interacted within that environment. Would you see the thrusters on a cloaked ship in space?

It's the same concept of the rock in the pond, but in a different environment. It's also possible to suppose an intelligent society has figured out a way to cloak any environmental interactions, but we can suppose not all intelligent societies have the same level of technology.

After writing the above paragraph I came across a critical article a few weeks later about cloaking effects near water. Jacob Dirhuber is an author for The Sun who wrote about a Navy incident in 2004 through leaked reports from the U.S. Department of Defense. In the article the UFO tracked the USS Princeton for a few days eluding contact by F-18 Hornets trying to intercept the alien craft. A specific reference in the article states "It appeared again two days later, and a pair of high-tech F-18 jets were scrambled to intercept it, but pilots reported that the object had turned itself invisible. It could still be detected as it was triggering a circular disturbance in the water "about 50 to 100 meters in diameter". This is a perfect example of in situ data collection proving a theory through credible eyewitness reports. In this case, the shape of the spacecraft was seen only because it interacted with the water around it.

Science is based on observations of our surroundings of both the animate and inanimate. It is why science fails to explain the origins or behaviors of those unseen forces. If an advanced society wishes to remain hidden while exploring the universe it is possible, we would not be able to detect them with the limited remote sensors we possess, or small frequency band of visual capability found within our own eyes. Therefore, we would have no direct observable evidence of highly advanced Extraterrestrials and especially if they did not want us to see them. My theory and the reason I am writing this book is to point out not everyone is on the same technological plane or path, to include Extraterrestrials, and some of them can be detected if one knows what to look for.

Decades ago, Exoplanets were theorized to exist simply by using our solar system as a model. Eventually, Exoplanets were detected using light-dimming detection of stars having planetary orbits crossing the line of sight of our star-gazing sensors. Recently, Exoplanets have been directly observed. Many peer-reviewed articles have not only shown we can detect Exoplanets, but theorize every star contains 1.1 planets orbiting the star that can sustain life in some form or fashion.

Gestalt & Imagery Analysis Principals

For those fortunate enough to have decent vision, scientists pondered the question of just how do we visually navigate our environment? We all take "seeing" for granted since birth, but science is still trying to figure out just how we see what we see. There are two aspects to vision: 1) the processing of environmental stimulus such as light/frequencies and 2) the cognitive process of making sense of what we see. These are two distinct processes that are sequentially handled when your eyes are open, but what if they are not? This is when your cognitive process kicks in to use other senses or produce dreams or can be directed by visualization.

A group of German psychologists in the 1920's decided to document basic principles we all cognitively use to determine what is in our immediate environment. Their work laid the foundations that are still cited in professional journals to this day. They include Proximity, Closure, Continuity, Similarity, Figure & Ground and Parallelism.

Proximity is a concept where multiple objects are close enough to or in a familiar formation (intelligent organization) to assign a relationship to the objects. However, the elements do not necessarily need to be consistent in shape to acquire this property, but location relevant to each other is the core of this principle. Continuity is another Gestalt principle where elements are arranged in a successive manner and linked together cognitively leading the observer to connect the elements in a smooth, relational manner. By observing image elements in succession strengthens the concept of continuity or outline of an object within the image. Figure & Ground is a principle where a common figure is recognized due to the arrangement of the background elements. The brain needs to interpret the patterns in your eye in terms of external objects or stimulus. This is accomplished by distinguishing objects (figure) from their background (ground). Another important Gestalt principle when dealing with space-based imagery is closure.

Closure presents element information that is incomplete in some manner like an open loop, but the mind tends to associate the connection based on prior information or experience. When human cognition compensates for missing data, it tends to close gaps of information that makes sense to the individual. The principle of symmetry involves elements located near each other forming a group in our mind and spatially related. The group does not have to contain elements of the exact same shape but can still be grouped together like the sections of a kite. Relative size plays an important role as do similar lines between the pieces. The Gestalt principle of parallelism refers to similar shapes or lines that are perceived to be duplicates of each other and therefore related or grouped. When similar lines are stacked near each other they are easily grouped or given a relationship. When these same lines are then moved in proximity to other lines not related in shape or size to the original group of lines strengthens its relationship due to parallelism. The final Gestalt principle used in this paper is similarity. When elements have similar color, size, and orientation tend to be cognitively grouped together. Even when one or more of these characteristics are missing it is still possible to group them using similarity.

Beyond the 1920's, the military application of these principles was applied when cameras were used to collect battlefield intelligence. The imagery created would then be analyzed for critical decision making throughout the course of wars and conflicts to this day. Basic compare and contrast techniques became the foundation and training of imagery analysts and include: Image Tone – contrasting light and darkness, Texture – apparent roughness or smoothness, Shadow – length of shadow, Pattern – similar lines or elements within an image, Association – similar objects tend to be grouped together, Shape – outline of the object, Size – in relation to environmental scene, and Site – topographical location or where the object was found regionally. Understanding both Gestalt principals and imagery analysis techniques it is not hard to apply these basic principles to any image regardless of what the camera or sensor scene has produced.

Remote Sensors

Remote sensors, by definition, are any apparatus used to detect the environment around it and transmit the results to another location for collection. The oldest recorded remote sensor was the Centurion. As long as sovereign societies and militaries have existed, Centuries and Spies were used to collect information for the commander to use in their decision-making process. Even a mirror on a stick overlooking a wall can be considered a type of remote sensor. I am asked all the time "why do I use solar satellite images"? Like a number of major discoveries, the thing discovered was not the thing sought for, but by accident. My discovery was no different. As explained earlier in this book, I was simply looking to see if I needed an umbrella for the day. I used a remote sensor called weather radar to determine the amount of precipitation or rain over Daytona Beach Florida prior to going to classes Embry-Riddle AU. When I saw the 4 or 5 unusually large ring patterns form and then dissipate exactly at the same time, over the whole state I decided to use another remote sensor to look for explanations. This time I used the space weather sensors that happened to be pointing at the sun. During the 24-hour time loop of the solar sensor I saw 88 images in rapid succession showing the latest solar activity. Within those images flashing in quick sequence, I saw strange shapes appear in one or two of the images. When slowed down to look at each individual frame, one or two images displayed unnatural shapes (lines of brightened pixels) not expected in deep space. Even though I found the answer I was looking for, no need for an umbrella, I could not help myself in chasing down the other question concerning the strange weather pattern and now these strange space image patterns.

My classes for an Unmanned Aircraft Systems degree included remote sensors as they apply to drones, but in fact, it applies to everything that gathers data remotely. I had collected a good number of NASA images from my favorite three satellite web sites and shared them with Dr. Haritos, one of my professors teaching remote sensors at that time at Embry-Riddle. He looked at the images I had offered and immediately asked if I had researched all the possibilities for image artifacts that degrade or interfere with image data collection. I told him I had not, and he advised I do further research to determine if what I was seeing in these images were documented artifacts created by cosmic rays, negative sensor characteristics or data post-production errors such as image compression. Dr. Haritos wisely required I find other possibilities for these weird objects and patterns to ensure I had covered all bases before making unusual claims of extraterrestrial photographic evidence. It was the process of research he emphasized to ensure we did not skip a step that would reveal a different conclusion. Dr. Haritos was a champion of "Compare and Contrast" research, and asked we submit a paper using this technique to report on a subject involving remote sensors we were interested in researching. Of course, I jumped at the chance to prove what I was seeing was exactly what I was claiming,,,something off-world that was intelligent.

What I discovered was some of the images and the weird artifacts I found were explainable through peer-reviewed literature and original sources from those who created the sensors themselves. Much like a 35mm camera creates lens flare (or ghosting) when pointed toward a bright light source, some of the satellite sensors had similar negative characteristics that show up in the image when certain environmental conditions existed. The charged couple device (CCD) itself contains pixels or micro sensors that measure bands or specific levels of frequency intensity between 0 to 255 increments of intensity or full charge. When processed and duplicated on another screen the scene that created the different patterns of intensity show up as a replicant of the image that made it. That image theoretically should be pure in replication, but other factors begin to creep in to degrade the data such as over-saturation of the pixels, lens distortions based on frequency and frequency intensity, "Cosmic Rays" that are a generic catch all for multiple types of high energy atomic and sub-atomic particles. Elevated temperatures can make CCDs inefficient in processing data and smear high intensity pixel levels into subsequent or follow on cells adding additional charge to the pixels being processed behind it. This smearing effect is called CTE or Charge Transfer Efficiency and will be discussed in greater detail later. It degrades the data along with all the other destructive alterations that over-write the original scene.

Some of these artifacts can be mistaken for pure image information or something real in the scene. Second generation decay of neutrons produce muons that can penetrate large swaths of material like earth by several 100s of feet. These muons are the size of electrons but 200+times as dense. They create higher levels of energy when traveling through pixels that show up on the image as a bright streak or dot depending on the angle of trajectory. By comparison, muons are the size of a BB flying through a small town when they cut through a single pixel measuring 15μm. These sub-atomic particles can strike multiple pixels in succession but not simultaneously such as two or more pixels side-by-side. However, the pixels have charge drains that allow a small transfer of energy to the pixel next to it as a cascading effect. Even this does not reach more than two or three pixels away from the original pixel that was struck. When dealing with high resolution CCDs or 4K architecture, a row of pixels being affected by cosmic rays is usually one pixel in width. In other words, extremely tiny.

As I learned the ways a single CCD image could be damaged by environmental factors, I could recognize the defects faster and exclude these anomalies from further research. This was the goal of my professor all along as he, too, walked down this same road of discovery in his studies regarding remote sensors. Discovery of something new in its pure form has already excluded all the possibilities of what it is not. It is from this position of uniqueness we can continue the scientific process of testing to determine the characteristics of what has been discovered. My paper for this remote sensor class was designed to address these destructive artifacts within CCD image production and then compare it to the items within the image that remained unexplained. The paper I wrote was a typical "Compare & Contrast" submission that helped me to better focus on those objects worthy of further research. I did have a flashback moment when looking at x-flares and how they force the satellite sensor to create a streak of light from the energy source on the surface of the sun to the edge of the image frame. A few months back I made the amateur error of thinking this was a laser beam attack on our sun from an alien source and reported it to a friend who didn't know any better either. We were both in shock until I corrected myself in a humbled follow up email confessing the magnitude of my error. Yikes, that was a hard lesson to learn. I reserve my findings now to those who know more than I in this field, however, I still mess with my friends just the same when I can get away with it.

It turns out there are a lot of reasons images from space sensors get corrupted. They range from the list below:

-**Beacon data:** It is the data initially transmitted without going through the onboard processing unit sent to a variety of collection locations on earth for further processing. It is very low quality and requires several collection systems to gather enough data to make a complete image. The science data is collected from the sensors onboard and processed onboard into a much higher resolution image before being transmitted to earth through the NASA Deep Space Network already in place.

-**Stars, planets, and comets:** These are the natural objects expected to be seen in space imagery and are not considered artifacts. However, to the untrained observer mistakes can be made as to what exactly is represented in the image.

-**Background subtraction:** Multiple images are combined into one depending on the sensor type and require computer calculations to combine them into an image that is better optimized. Some of these calculations denature the image too far and produce dark regions not a part of the original image.

-Cosmic rays: It is enough to say cosmic rays are a generic term for a whole host of highly active, charges particles and frequencies that degrade an image by impacting the CCD architecture during its operations. Most of the particles that strike the sensor CCD are millions of times smaller than a single pixel on the CCD but carry enough energy to increase the charge levels artificially. These charges are then added to what is naturally collected by the pixel during exposure and then displayed as straight-line streaks or dots depending on the angle of the impact.

-Debris: This term encompasses anything man-made detected by the sensor or not naturally occurring. Some of the debris can be from the sensor protection blankets when struck by micrometeorites and cross into the sensors FOV as described on the NASA STEREO website.

-Camera defects: Those familiar with photography know lens flares or ghosts are naturally created by the crystal lattice structure of the glass lens acting with light through reflection and diffraction. Lens materials vary but can create artifacts depending on light angles and intensities. Space based solar satellites are no different and have the same types of lens artifacts associated with them.

-Spacecraft rolls: Satellites are sometimes physically rolled by command to recalibrate onboard inertial systems. During these roll operations images are still produced but only partially and with unusual results.

-Internal reflections: The lens edge and barrel reflections can also change the light characteristics within the image depending on intensity and source location. It can produce lens flares like artifacts that degrade image quality.

-Corrupted and blank images: Data transfer can be interrupted or degraded to a point of incomplete image creation. A major characteristic of this are blank portions of missing data or large pixelated sections that offer little to no resolution for qualitative analysis.

-Charge Transfer Effect: This artifact is very complicated and rare. It deals with not only the physical structure and material stability, but also the digital processing techniques and sequencing of data. It is also the most spectacular to see as it follows objects creating a high intensity source within the image and creates what looks like an exhaust trail.

Image compression is done to satellite pictures to reduce the data amount to its lowest level while maintaining the highest quality of the original data set. This allows more images to be taken in less time and less data transmitted in total. The effects can create what appears to be pixelated or artificial structures within the image causing misidentification of naturally occurring cosmic rays strikes through the CCD structure. I know this to be true from my own experiences as I was sure I found alien spaceships all over the solar images in the NASA archives when I first started my investigations, but I was wrong for about a year reviewing thousands of images until one day I was right. It took me a long time to realize just how many artifacts can be found on one image that are nothing more than picture defects due to the highly active and charged space environment.

CCD Architecture

Tech alert! Please do not read this section if you have no desire to learn the design characteristics of Charge Couple Devices. The only reason I write about CCDs is because it is extremely important to debunk a claim by astronomers who know a lot about stars but not a lot about electronics or CCD architecture. The current trend in the astronomy community that all images containing bright artifacts followed by a long trail near the sun in a single frame is caused by CTE is not accurate. CTE is an effect that encompasses the whole shift register within the CCD architecture and smears all pixel charges at the same rate given the charges are similar in intensity. The smear effect is evidencing the shift register circuitry can't discharge the previous pixel charge all the way to zero-charge before accepting the next charge cycle and sending it downstream for processing. This effect leaves a smooth streak or trail behind the pixels with enough intensity to trigger the effect. Once the register can reach a zero-charge state, then the smear ends until another triggering charge reaches the register. This means if a streak starts and then is broken (reset to zero), it will stay zero until another pixel of charge is received. Therefore, if an image of an object creating a streak has broken line segments, then the trail is a real physical event (like a comet tail) and not CTE or something created by CCD material instability. (See Fig 1 – page 36).

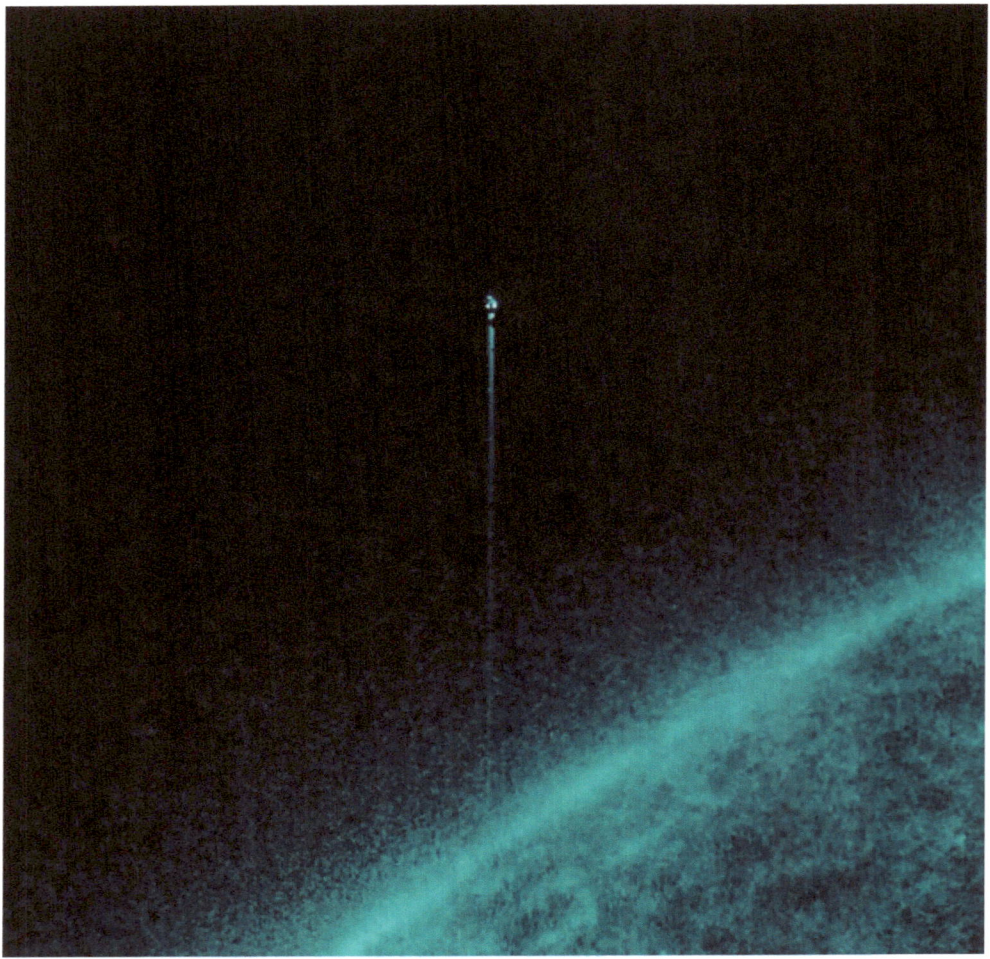

Fig 1. Object with trail behind the sun. (Courtesy of NASA: SDO satellite date/time/size/sensor group - 20180510_125832_4096_0131)

The focus of this section was technical by nature, but necessary to refute some of the facts offered by NASA as theories. Not all information offered by NASA is vetted to a point of fact, but concepts like CTE do hold enough probability to reach theory status for some objects. The real issue is when every object in solar imagery that has an apparent smear behind it is automatically labeled a product of CTE, then we miss discoveries covered by this blanket assumption.

Charge Transfer Efficiency is best described as CCD inefficiencies handling data transfer from the plate of pixels to the shift register immediately after the exposure is completed. When temperatures get too high for whatever reason the CCD materials handling the charges become slower discharging each cycle to zero before the next charge is read. This means as the pixels get downloaded row by row into the shift registers, the charge levels can't clear before the next pixel charge is processed thereby adding residual energy to the next pixel charge. A lab created example of CTE is seen in Fig 2 (page 37) where the image was taken in a dark room or with the shutter closed to allow cosmic rays to affect the CCD. The CCD was also put under CTE conditions to produce the smearing effect. The top image shows normal image processing, where the bottom image shows the effects of CTE as highlighted in the red boxes.

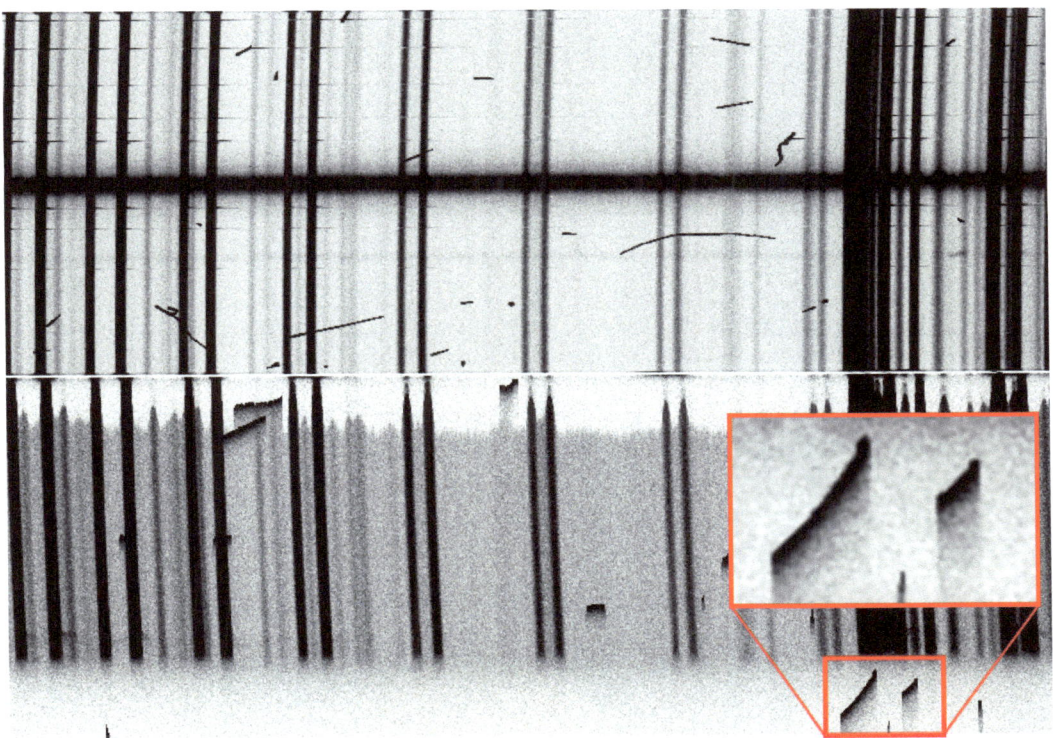

Fig 2. A 1980 pixel × 800-pixel subfield of a 3600 s dark exposure (totally depleted 270-μm thick LBNL CCD, NOAO CCD laboratory in Tucson), showing cosmic-ray muons (straight tracks), worms (low-energy electrons), and spots. While the spots look insignificant, they are about as abundant as the worm shapes and can indicate considerable deposited energy.

The CCD diagram below is the one offered by MIT Lincoln Laboratories who manufactured the SDO satellite systems CCD sensors. Each sensor is four CCD sections combined into one as seen in Fig 3 below. If you imagine the head of each Black arrow is an object with high intensity levels being processed, then the output on a remote screen (like your computer) would display the exact pattern of the arrow of smearing if CTE existed for this CCD. It is physically impossible for the smear to happen in the reverse direction for these quadrants.

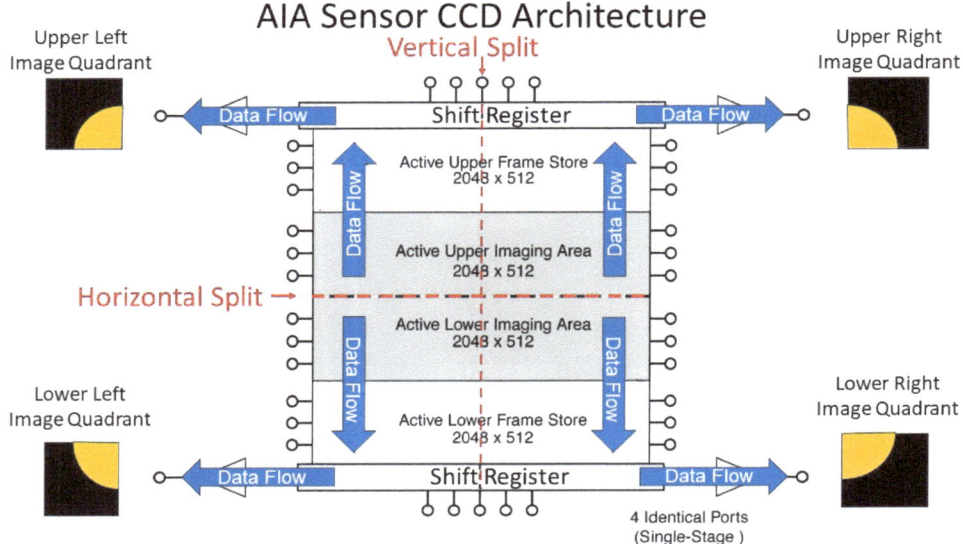

Fig 3. The above figure shows four CCDs fused together (grey area) along the dotted line with the dark blue arrows indicating the direction of data flow from imaging area to frame storage spaces above and below. From the frame storage the data moves to the shift registers that pull data out and away from the imaging area in a single direction only.

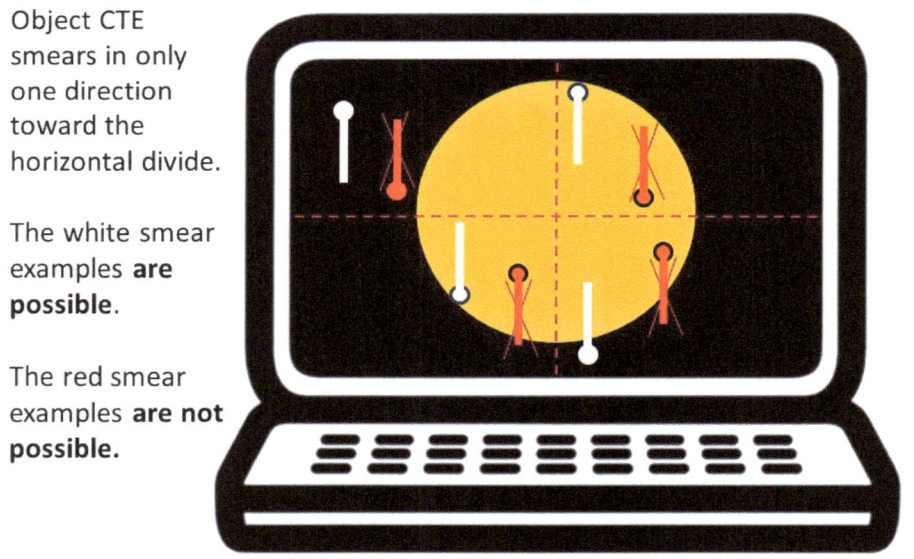

Fig 4. The computer display above shows the four image quadrants stitched together. A CTE (Charge Transfer Effect) can only smear excess charges within a line of pixels in one direction toward the horizontal center line as seen in the white examples based on shift register placement.

Each of the four CCD sections show a dark blue line indicating the data flow from the gray section of the initial picture exposure to the white sections where the image storage area is data sliced next. Figure 5 (page 40) below shows first-hand the four image quadrants making up the CCD sensor by the faint lines cutting the image into quarters. This image flaw is not normally produced but when poor image stitching occurs the computer program sometimes displays the four quadrants that can be seen individually. These quadrants are the result of the four ports that process the image data away from the CCD center, in four different directions. Calibrating the system and post-processing adjustments eliminate these lines when the system is running optimally.

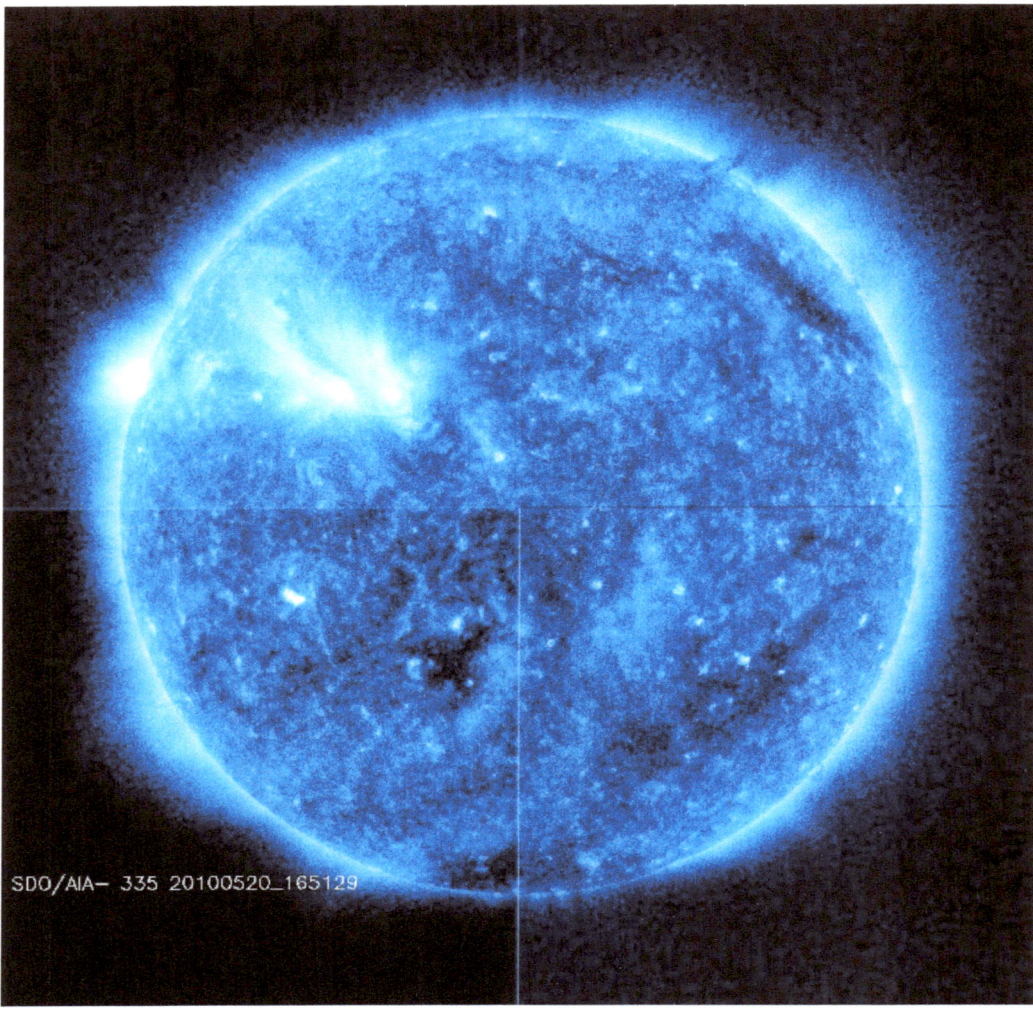

Fig 5. From four single CCDs within the sensor, these quadrants are digitally stitched together to represent the intended target subject (the sun) as seen in this figure. (Courtesy of NASA: SDO satellite date/time/size/sensor group - 20100520_165129_4096_0335)

If just one image can be offered as proof all smearing is not CTE related, then one must refuse the assumption all smears are caused by CTEs. Please review the image below as one of many proofs I have on file showing smearing in opposite directions within the same image quadrant by two different objects. This makes CTE physically impossible for one or the other objects as the shift register <u>can-not</u> "Pre-smear" an object before it reaches the register (See Fig 18, 19, 20 (page 41) – each figure contains two solar centric objects in the image frame moving in opposite directions).

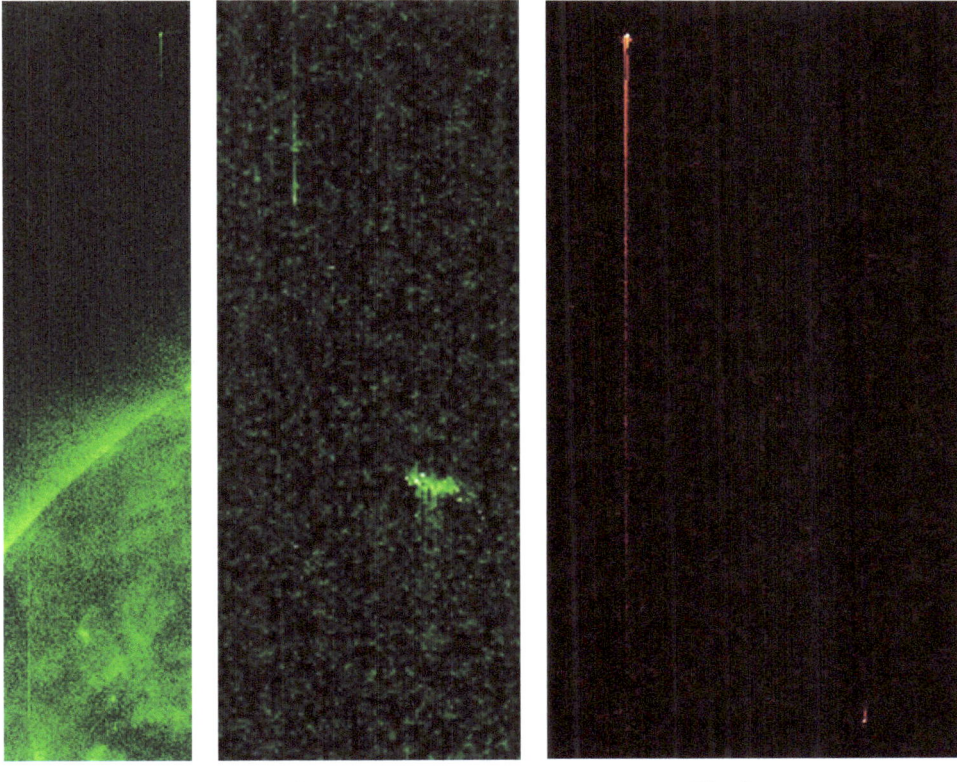

Fig. 6   Fig. 7   Fig. 8

Fig 6 - 8. Each of the figures above contains two objects within each image. More importantly, the smear directions are opposite of each other which is physically impossible for CTE within the same CCD quadrant. (Courtesy of NASA: SDO satellite)

These images may not show the detail of the split smear directions on paper as well as it does when digitally view from the original photos on the NASA website. However, they are proof assumptions can make you a poor scientist. In figure 6 above, if CTE created the smears or streaks behind the objects seen, then the rest of the sun would be smeared as well. It is not. Figures 6, 7, and 8 all have two objects with smears in opposite directions. CTEs can only smear data in a single direction within the image area. These examples prove CTE is not the explanation for all smears. If all the trails, smears and streaks were erased from the images the larger question remains: What are the objects creating if these are not the result of CTE? A possible conclusion for this effect is the object is creating a plasma trail due to energy transfer from the object on the surrounding aether or plasma field.

Each satellite has multiple Atmospheric Imaging Assembly (AIA) and magnetic sensors adjusted to specific frequency collection at different angstroms. Those angstrom lengths are listed in Fig 9 and associated with the objects with apparent trails. Most astronomers dismiss the trails as sensor created artifacts of CTE. I have just offered proof this assumption is incorrect.

AIA Detector Frequency

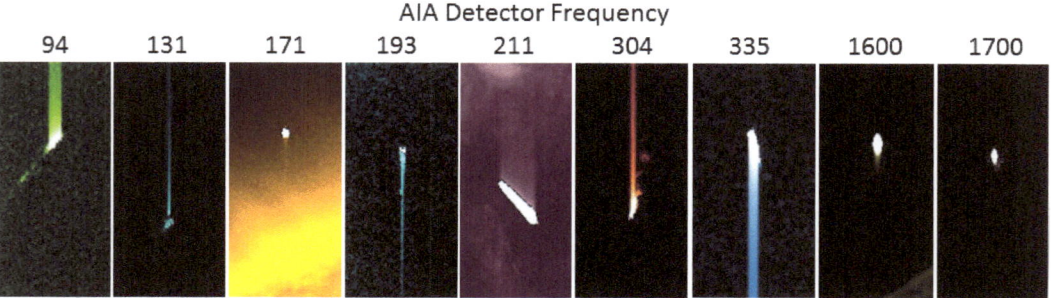

Fig 9. This panel of solar centric objects shows each sensor on solar satellites spanning all frequencies can pick up objects near the sun that are not naturally occurring objects (i.e., Comets, asteroids, debris).

However, the unreported characteristics are CTE requires a pixel or region to reach charge saturation (in computer terms level 255) or close to it before the residual charge begins to be carried over to the next pixel or region. If a pixel or region of pixels do not reach a high enough charge level when the conditions for CTE are present, then there will be little to no CTE smearing. If it did, then the sun would also be one large smear and produce worthless data, but some very beautiful artwork as seen in fig 10 (page 43) below.

Fig 10. This image is not the result of CTE happening within the lower shift register because the smear is away from the horizontal split of quadrants.

The other unspoken CTE characteristic is it does not taper to a point from the origin to the edge of the frame. Tapering happens when the environment being photographed captures a real trail or tail being produced like in a comet. Tapering is an indicator that the object or event is real and not an artifact produced by CTE. A prime example of a true trail produced by an object is seen in Fig 11 (page 44).

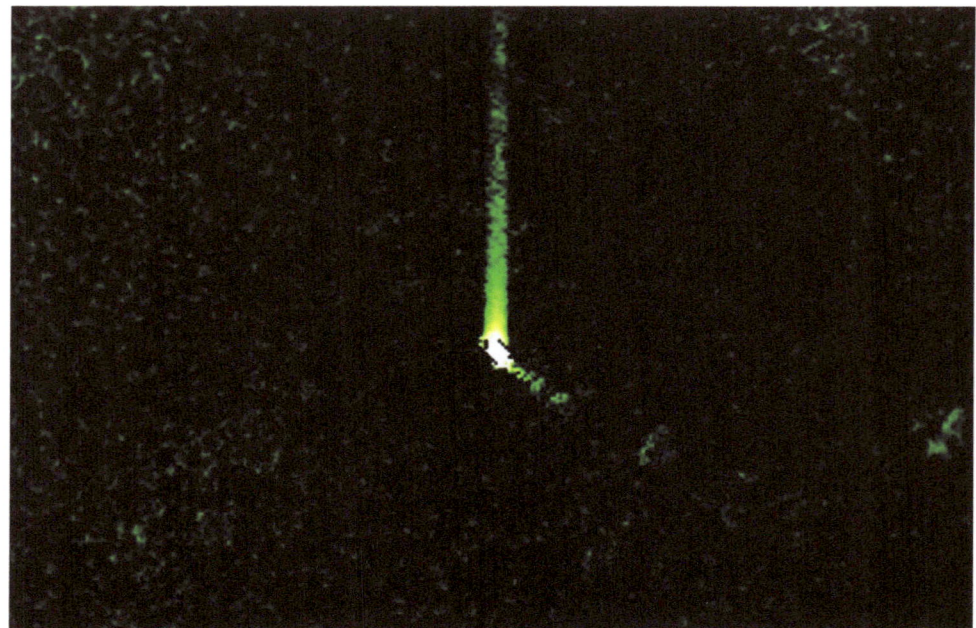

Fig 11. The object above shows the trail tapering into a point not seen within a true CTE event.

CTE is consistent in smearing once the process has started. It may fade with time as indicated by length of smear, but once the smearing ends or the pixel cells are back to zero in the shift register, then everything is reset to proper levels. No more smearing should occur after this point is reached by the original pixel or region of pixels that initiated the CTE. However, if the trail is real, then it is possible to have portions of the trail blocked or processed out of the picture leaving the remnant of the trail to be photographed (See Fig 12 - page 45).

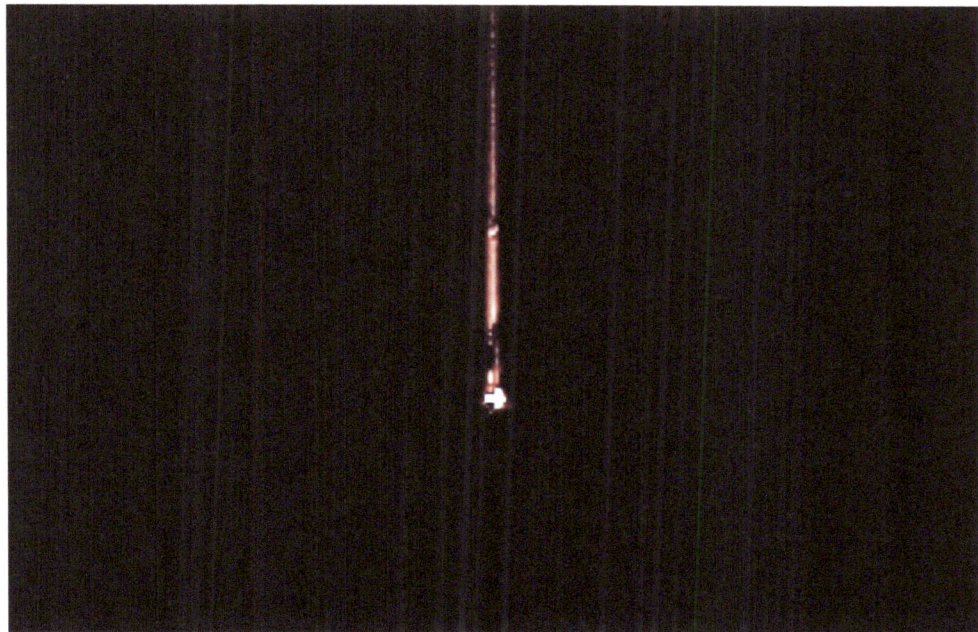

Fig 12. There are breaks in the trail behind the object that cannot be created through a CTE condition due to this quadrant can only produce smears in the opposite direction.

Notice all the breaks in the trail or zero points of charge, and then it reappears. CTE does not randomly add charge to discharged pixels once they reach zero or recreate a smear once the charge is zeroed out in the shift register.

Theories and Facts

In the NASA STEREO website there exists a page exploring the image artifacts that describes debris as a potential reason for unusual objects crossing the FOV and reflecting the sunlight back into the lens. The website claims the debris is from micrometeorite impacts on the satellite protective insulation blankets. The first question I have is how does the material reflect directly backwards into the lens (see Fig 9)? As a side note, this image is no longer available on the same website using it as an example. Another curiosity is the particles they claim blown off the protective blankets by micrometeorites are all the same shape and size. Materials ripped by a high impact event should exhibit irregular shaped pieces leaving the impact sight. This is not what you see in this image. It's okay to question what we are told by those we are taught to respect, not as a form of rebellion but one of discovery. NASA has too often been silent on certain videos or images that are difficult to describe accurately without offering Extraterrestrial possibilities. Even if it is not what it is, then why not at least throw it in with other possible answers. You be the judge as to why.

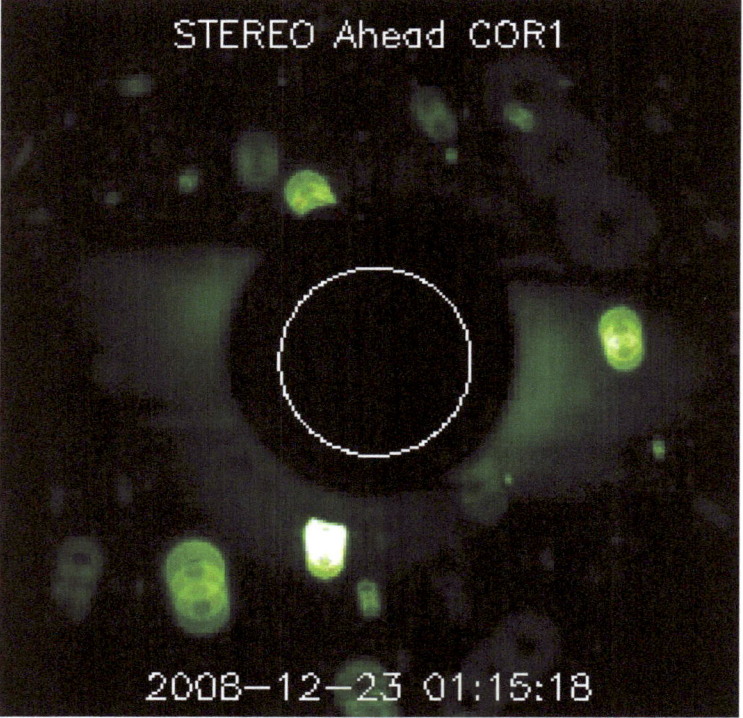

Fig 13. NASA image used to support the claim the lighted areas are structural pieces of debris from the satellite being struck by something crossing its path.

NASA uses the image in Fig 13 (page 46) from the STEREO Ahead COR1 sensor as an alleged example of floating debris from impacts against the satellite. However, since these items only show up on one frame, the speed of the debris seems to be extremely fast. The cadence or time between exposures in theCOR2 sensor archive is 15 minutes. This means there should be a cloud of impact debris surrounding the satellite as it travels in the following frames. There is nothing like that indicated. They offer up a second image from the COR2 sensor next to the COR1 and quietly imply the same impact event is happening in this image too (see Fig 14). The two images are nothing alike by object characteristic or signature of shape. The only thing they have in common is they appear only once in a rapid sequence (or cadence) of images in 99% of the cases.

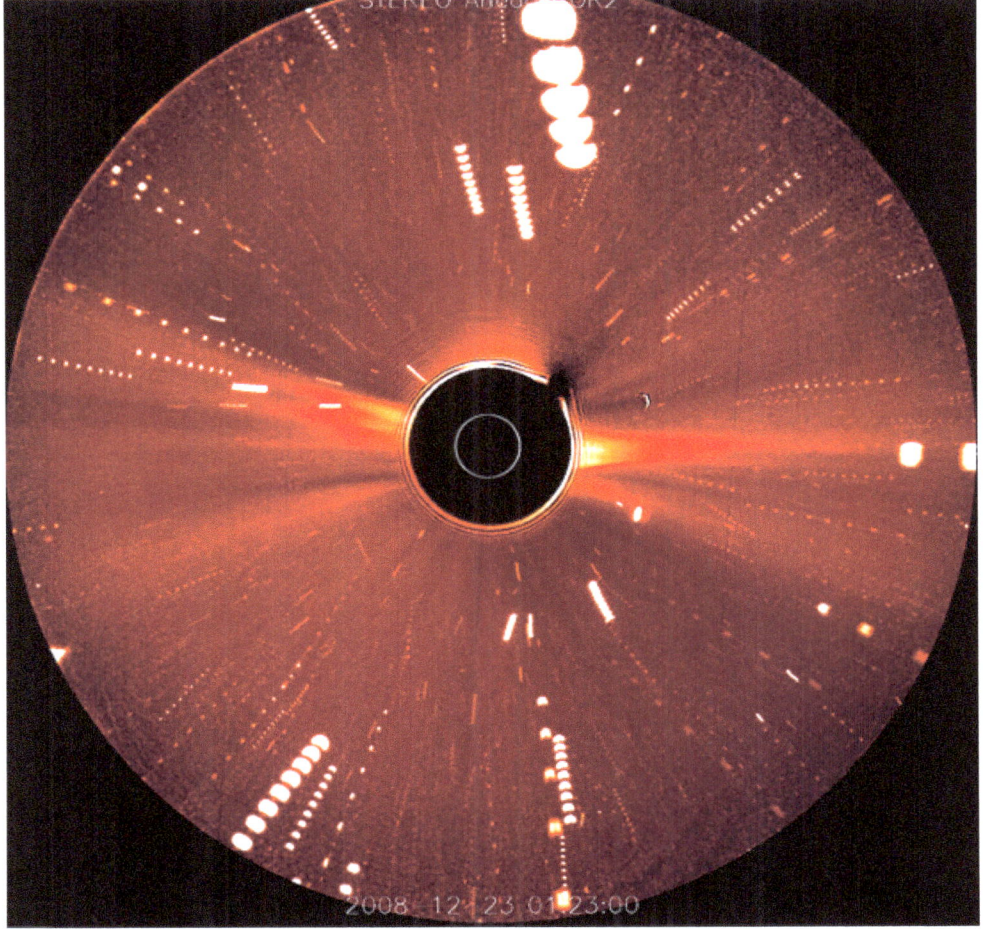

Fig 14. Image is from a COR 2 STEREO-A sensor.

Micrometeorite impact is an unsubstantiated theory offered in literature for these image artifacts (Brown, 2018). If you had that much debris flying off the satellite, the debris would hardly be traveling in a straight line towards the sun and especially since the debris seems to come from multiple location surrounding the satellite. If struck from the top-left portion of the sensor area, debris would travel in multiple directions from a single location and then cross the FOV showing the same. I offer Fig 15 below as an example of cross FOV trajectories of alleged debris, but the problem again are the objects seem to move in parallel trajectories to each other (mostly), and not from a single point of impact displaying a point of origin associated with the satellite. Also, the distant sequence of lights indicates their origin is nowhere near the satellite yet have a cross-trajectory pattern (parallel to the movement of the satellite with apparent speed greater than the satellite).

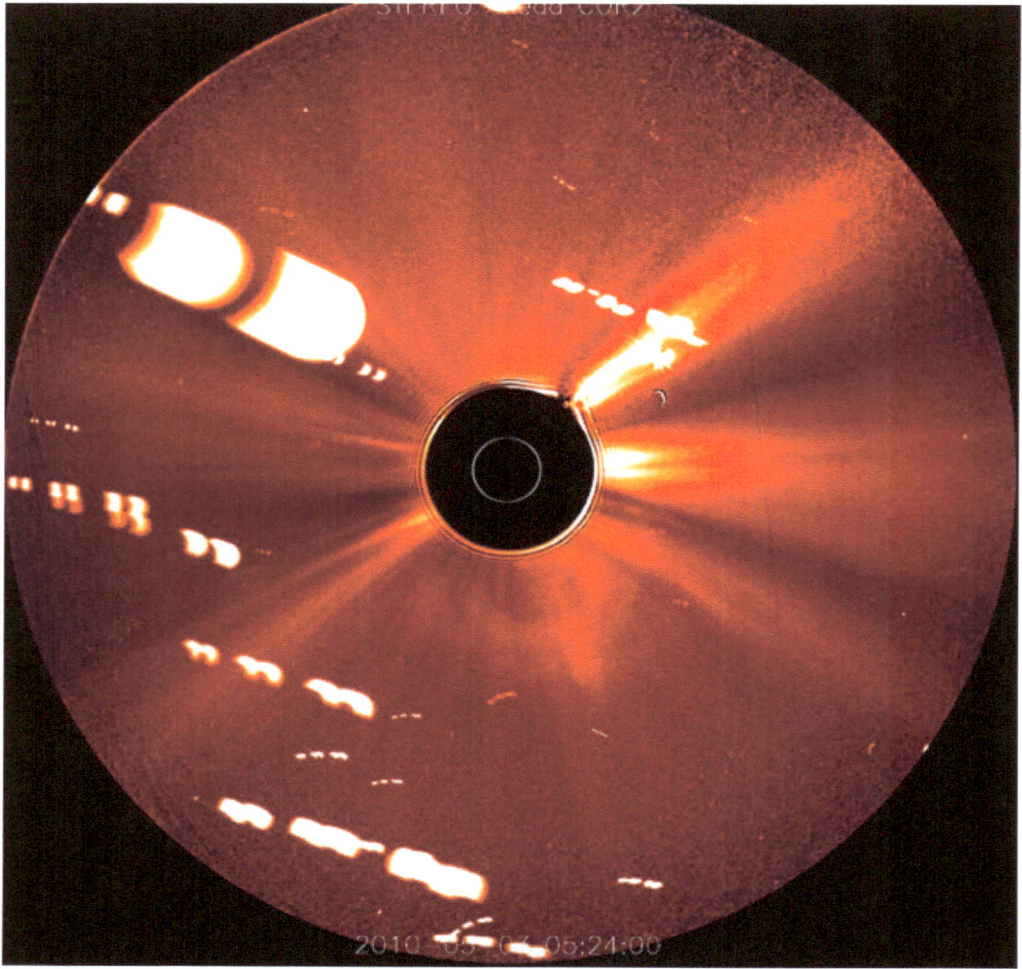

Fig 15. The image is the STEREO-A C2 sensor with objects crossing at different angles.

The image they choose does not show characteristics consistent with the claim. The image above likewise debunks this theory of micrometeorite impact due to obvious dissimilar trajectories, sizes (difference in distances), and a speed component greater but equal with each debris particle in the same direction. In both cases the speed of debris departure is incredibly high as there is little to no evidence of a debris cloud remaining in sequential images below.

COR2 10/5/2003   COR2 12/23/2008

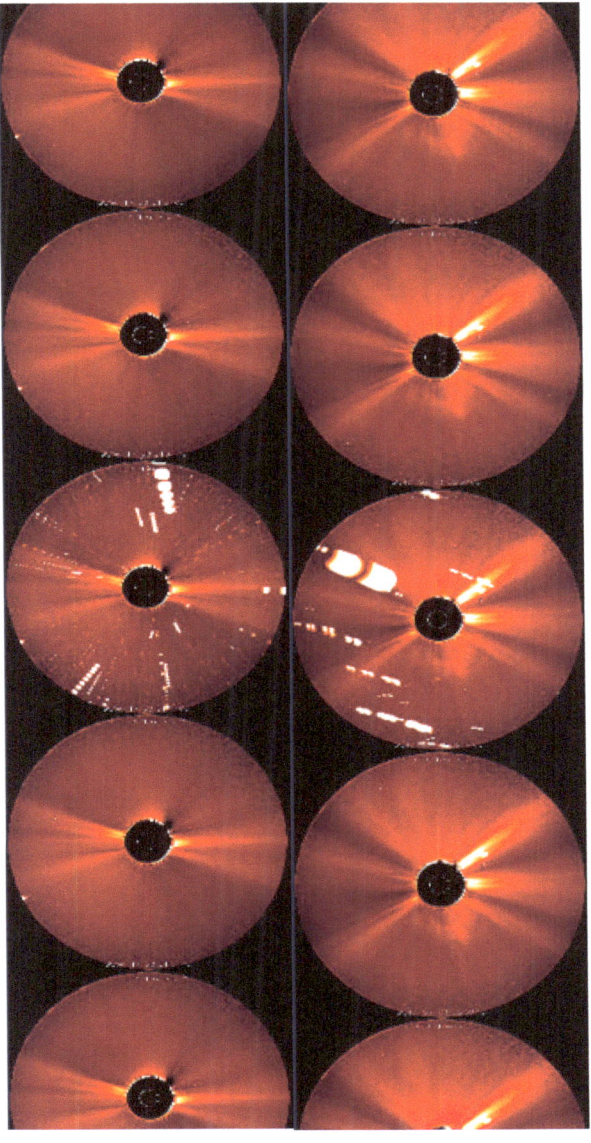

Fig 16   Fig 17

Fig 16 - 17. Each of the figures above contains a series of sequential images where one image in the vertical series contains multiple objects. (Courtesy of NASA: SDO satellite)

It is my analysis that these same objects have other qualities that indicate controlled flight as similar objects show changes of trajectory near the satellite. In the series of images below similar objects display a variety of flight paths that require an external or internal force to change direction as shown in the photos below. Some images show multiple object trajectory changes in different directions within the same frame/figure (see Fig 18, 19, 20, 21).

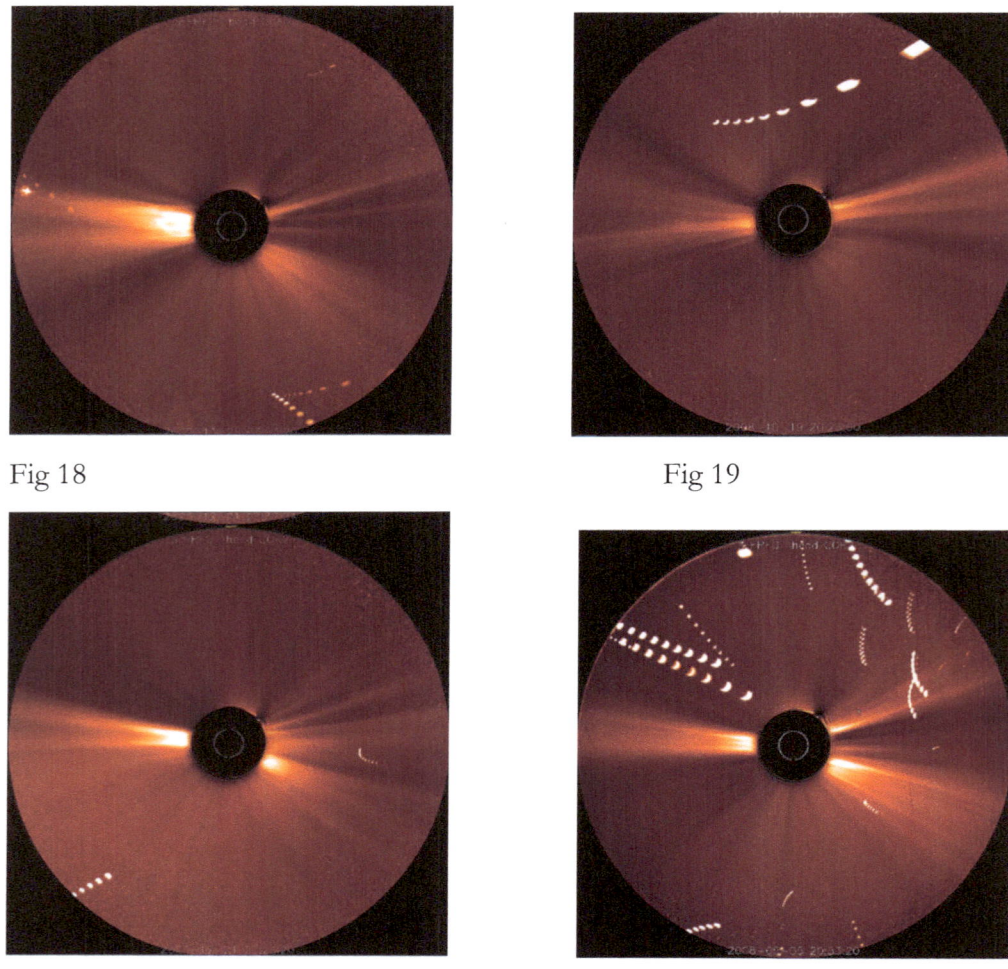

Fig 18

Fig 19

Fig 20

Fig 21

Fig 18 - 21. Each of the figures above contains one or more objects indicating intelligent trajectory avoidance of the satellite/sensor. In Fig 20., multiple objects display random trajectories indicating either multiple initiating sources converging in front of the satellite sensor by chance or intelligent control is involved. (Courtesy of NASA: SDO satellite)

I can offer two visual rebuttals to their claim; 1. No particle cloud is present in the following image frames and 2. All the claimed particles are the same shape which is impossible. I can't remember a single explosion in my entire Army career, explosions I have witnessed or ones I have created, that produce identical shapes from the debris. The only way to achieve their claim is to blow up a BB factory. Even then there would be BB container shrapnel of various sizes mixed in with the BBs. It doesn't hold enough reasonableness to exist as the only explanation. There is a third rebuttal to the COR2 image offered by NASA and it is this; the pieces of debris saturate the CCD pixels to maximum charge equally. Why equally? Different-sized particles from an impact would offer various angles of light reflection thereby offering varied light intensity. The image shows the opposite has happened. The visual characteristics in this image show identical objects producing their own energy at a rate high enough to saturate the CCD pixels. What kind of objects would do this at 30 million miles away from the sun? A better theory based on what is displayed is that these objects are space drones from our planet or possibly from an extraterrestrial one. Direction is not easily determined in the COR2 data set, but the unusual and continued fly-by past this sensor makes the satellite look like a way point or reporting point for space travel. Perhaps the satellite doubles in duties as a remote Air Traffic Control system much like an airport uses.

Portraits in Chaos

This leads to the next big question: Is there a way to observe the objects' exterior? Every photographer knows how important lighting is to properly capture all the nuances of their subject. Lighting position, to include angle and intensity, makes the difference between poor photography to award winning photography. The obvious advantage of portraits or still art is the control the photographer has on the subject and environment. Every item is placed and adjusted to optimize the frame before the shutter is activated. Even so, the results are only realized in the final product. Wildlife photography is harder because many times the randomness of the subject, action, and lighting are all in flux at the time of the shot. However, most wildlife subjects are not traveling at speeds beyond the sensor's capability or photographer's skill set.

When considering the cameras on a satellite you have entered the world of capturing images and objects involving hyper-speeds from lightspeed on down to myriad of frequency ranges on both sides of our own ocular capabilities. Likewise, with regards to satellites you use state of the art sensors bolted together in banks of different flavors of frequencies, strapped to a rocket and shot into space at unheard of speeds. Add to this throwing a camera/sensor into a hostile environment to include exposure to damaging energy and particles from every angle and intensity known to man (and some still unknown). The photographer becomes a small group of remote sensor scientists who code into a program X, Y, Z coordinates and ask the camera to respond accordingly. Once this complex package of camera technology begins to function as designed, the subject of all the effort is still to be explored: in this case, the Sun. Getting the cameras/satellite under control is half the battle while gathering data is the other half.

As in any high energy environment, the solar satellites must maneuver into proper orbital trajectories, stabilize the platform for optimized image quality, conduct maintenance tests, and maintain data and telemetry communications. When all the system sensors operate as designed the result is image production of solar events catalogued and labeled by the Heliophysics Events Knowledge (HEK) database. This system is automated and captures a variety of events computer programs can determine as significant. Satellite sensors are designed to capture clean data, and because of this the events on or near the solar surface can be easily determined. Most of these events are common given the amount of repeated and discernable events from the past archives and clever computer detection programs. The HEK is an image repository including the physical attributes and events associated with the sun. However, there is no current repository of image data or computer programs for astrobiologists to catalogue Technosignatures.

Since Astrobiologists have decided to look for life on Exoplanets the limitation of current sensor resolution has hindered any direct discoveries. If we simply sent a satellite to a new solar system containing planets like our own, it would be easy to determine what kind of life existed there. Since that would take light-years of travel time this scenario is not likely. The other fix is creating such high-resolution space-based sensors as to properly see any evidence needed to determine if life existed there. The question is, what happens when life is detected, and what will that look like? On September 20th, 1973, the three Skylab astronauts (Bean, Garriott, and Lousma) had a tuff time describing the possible Extraterrestrial Vehicle (ETV) outside their window to ground control. The image below was one of two possible ETVs they photographed that were otherwise indescribable (Fig. 22).

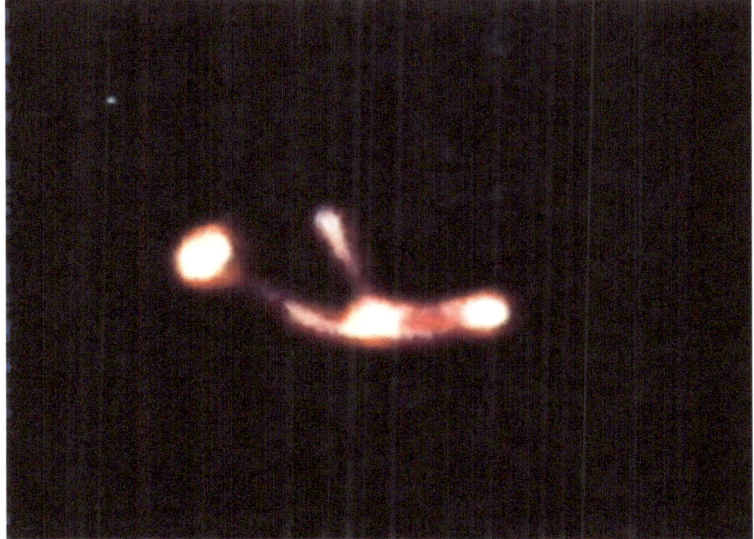

Fig. 22. Official NASA photo taken of an object outside Skylab on September 20th, 1973.

If this is a picture of an ETV or some sort of spacecraft, there is a whole universe of learning that awaits our species as humanity reaches out to find its neighbors. The obvious issue with this photograph is no one knows what this is to this day. The weird shape only lasted 10 minutes as reported by the astronauts Alan Bean, Owen Garriott, and Jack Lousma. Using this NASA confirmed data we can begin to see what challenges are ahead for Astrobiologists when more data like this emerges in the future. Without a solid start point of a verified EVT or signs of extraterrestrial life astrobiologists are unable to begin the cataloging process.

Another more recent incident captured by NASA cameras involved the STS-75 mission on the Space Shuttle Columbia that was to carry the Tethered Satellite System flight (STS-75) into space for electrical experimentation in February 1996. During the deployment of the system (see Fig. 23) the tether separated from the shuttle and floated away from both the ship and earth. During the accidental payload separation, the cameras onboard, the shuttle remained pointed towards the tether for as long as it was observable. The unexpected result was video proof the tether was surrounded by unknown objects that not only looked like odd-shaped donuts but some of them pulsated and changed trajectories near the tether (see Fig. 24 & 25 – page 55)

Fig. 23 Space Shuttle mission STS-75 Tether Experiment (Courtesy of NASA)

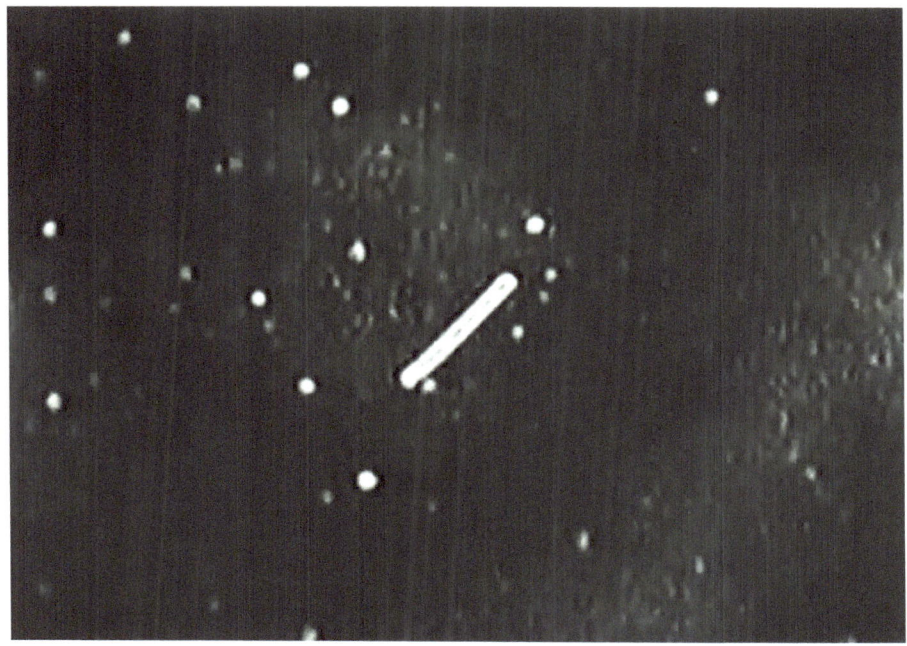

Fig. 24. Long object is 12 miles of wire inside the coiled tether. Image credit to LunaCognita found on YouTube.

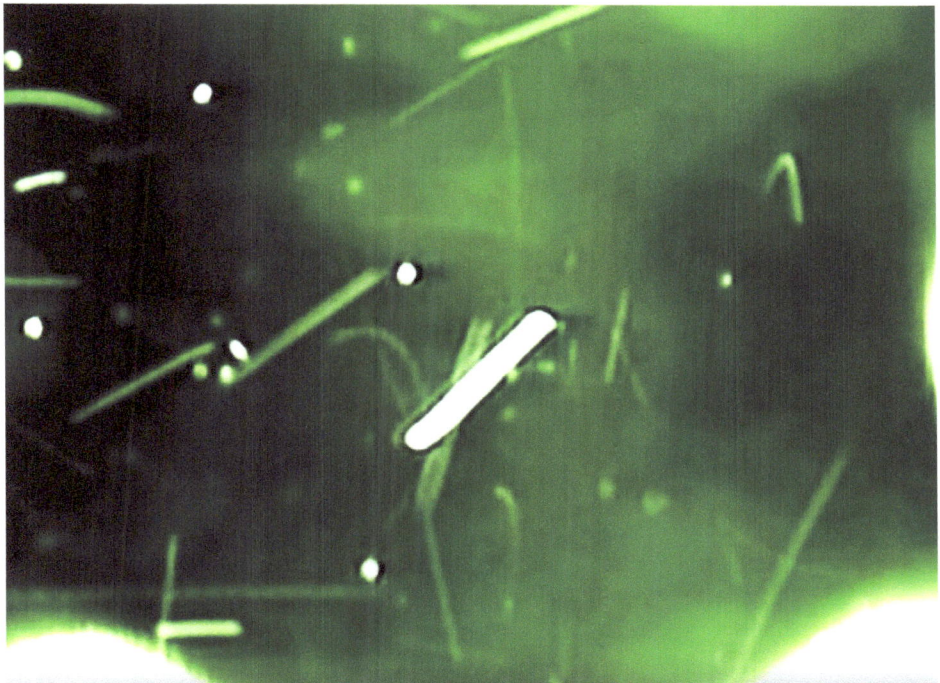

Fig. 25 Trail paths created by motion tracking program. Image credit to LunaCognita – YouTube video.

In figure 25 above (page 55) the bright white thick line in the center of the image is the tether that accidentally separated from the Space Shuttle during the STS-75 mission. All the other lines or dots are objects within the camera field of view. The green lines, however, are artificially generated as a tracking program designed to follow or leave a visible trail from the object creating it as it moves through the scene. You can see some of the "Trail" lines are curved indicating the object that created them has arbitrarily changed direction. This is only accomplished in this case by an intelligent force indicating control of trajectory. Nothing else explains why some objects move straight and others move in a nonlinear fashion.

Sir Isaac Newton came up with a few basic laws of physics that apply to the objects surrounding the tether in the pictures above. Objects in space, we are taught, move in a single direction for as long as no other forces acting upon them change that direction (Law of Inertia). This would imply every one of the objects seen near the tether would create a straight trail when subjected to tracking programs used to detect deflections or changes in trajectory. The objects in figure 25 (page 55) show changes in multiple object trajectories to include objects appearing and disappearing in the video of this event as if in a magic show. NASA has no explanation for these objects surrounding and conducting intelligent maneuvers. This second ETV example is offered as one of many unexplained incidents involving unknown entities under foreign control. NASA recorded these incidents while refusing to elaborate possible theories for this sighting that must include the possibility of extraterrestrial life as a viable option. The reasons for the silent denial by NASA only exacerbate conspiracy junkies to fill the void. It's hard to believe NASA supports efforts to find life elsewhere in the galaxy but refuse to admit extraterrestrial life may already be at our doorstep.

The main problem for Astrobiologists is identification. Detecting foreign entities or objects with intelligent characteristics has already been done. Collecting this same data through photographs and video has also been accomplished. Unfortunately, the incidents where ET life may be an explanation is rare and usually one of a kind. The ability to collect evidence that is easily identified as belonging to an already existing group of data was near impossible until recently. Unlike physics where math is used to autopsy materials or processes with a quantitative approach, ET life has been relegated to qualitative methods like imagery identification or frequency perturbations from deep space. Also, physics can stage the experiment by controlling the materials being tested where capturing objects in imagery relies on being at the right place at the right time and then positively identifying ET life forms as common. Since there are no catalogs of positively identified ET life forms, there is no baseline from which to work.

Given the chaotic environment of space and the intense physics that drives systems as large as galaxies the prospect of taming this environment to reveal the evidence of life being sought is slim to none. Using imagery from solar satellites has one advantage over other systems in that the sun is photographed like it was the subject of a portrait. These stable camera systems offer the HEK the ability to receive clean data to make measurements and discoveries that later become a part of the catalogue of knowledge about our sun. However, none of the categories listed in the HEK for solar events have anything to do with detecting intelligent life on or near the sun. WHAT? Life near the sun? That's crazy, right? Maybe not.

NASA's next solar satellite mission, the Parker Solar Probe, involves orbits of extreme closeness to the sun never before imagined at 3.9 million miles from the solar surface. The temperature will reach 2,500 degrees Fahrenheit at the probes closest point accelerating to nearly 1 million miles per hour. If humanity can achieve such goals from powered flight a mere 100 years ago, again I ask what might happen to other life forms who are thousands of years ahead of us? It is my belief the reverse of verifying life on other planets has happened, but to us. It's possible for thousands of years we have been the discovery for others to probe and verify we exist and are moving forward towards technological advancement to allow travel to other solar neighborhoods. It's also possible the bulk of our technology was gained by contact with these others who ventured to study us. Why is this concept so reviled by those who spend millions of dollars looking for life elsewhere?

Back to identification. When astronauts Alan Bean, Owen Garriott, and Jack Lousma reported the object outside their spacecraft, what shape was described (see Fig. 22 – page 53)? There is nothing consistent in geometry to describe a shape like this or many other NASA documented unknown objects since they are foreign to begin with. Qualitative analysis requires commonality based on a strong operating definition. It must have some sort of similarity to each other to build categories that can be archived for future research. This is why I show examples like Fig. 9 (page 42) to prove there are Technosignatures offering similar shapes and characteristics allowing a group or category to be formed. These objects not only appear to show maximum energy output near the COR2, but some appear to maneuver to avoid hitting the satellite. The sheer fact of trajectory change displayed near the sensor gives this object the intelligent trait of directional control. I offer this Technosignature as the first of its kind to be properly evaluated by the astrobiology community as a starting point for extraterrestrial discovery.

Expectations

NASA has cleverly encouraged exploration and discouraged explanation all at the same time. Only recently have scientists been supported in efforts to test for microbial life in outer space or discuss panspermia (life being distributed throughout space) as an acceptable theory of how life started for us. Why not let that theory mature in our brains for a second and extrapolate the next step in having progression of life throughout the expanse of space accelerate at different rates for just a few billion years? I believe Dr. Drake eloquently offered a formula for this very reason as discussed earlier in this book to set expectations. However, I believe a lot of smart people using Dr. Drake's equation in support of their theories failed to see the obvious missing piece; space itself.

When adding up all the potential planets for holding life within a galaxy, what percentage of mass/surface area is involved? I don't pretend to know as no one else can answer that question besides God. However, we do know it's astronomically less than 1 percent. That means Dr. Drake has focused our attention on something so small numerically, the odds are low of finding life elsewhere, but still greater than 0. If we are conservative with the probability of life with tens or thousands of exoplanets in our galaxy, then finding life on just one is not so crazy statistically. By offering the rest of space as a region containing life that is traveling, we expand Dr. Drake's equation by over 99% for possibilities. If I increased your odds at winning Blackjack in Las Vegas by that same amount, you would throw this book down and be on the next thing smoking to Vegas. This new look at where to look for intelligent life just grew by 99%. By the way, we can now look for life in our solar system now. Oh yeah, one more thing, we can use our solar satellites for dual purposes in astrophysics and astrobiology exploration. See what can happen when you set up proper expectations? You become free from the bondage of boxed in science and talk freely about the what ifs on the same level as any other theorist. More importantly we can start looking for other life forms closer to home.

Portraits of the Sun

The reason I use solar satellite imagery is simple, the subject is stabilized, the sensors are optimized, the data collected is reliable and valid for research purposes. However, the most important reason I use solar imagery is lighting. I alluded to this earlier in the book in the last chapter but let me expand on why it is so important. If a sensor is looking at only one frequency and an object is not creating nor reflecting that frequency, it remains hidden. If an object can camouflage itself, it remains hidden. If an object cloaks itself, it remains hidden. So, to "see" objects with remote sensors, we need to match the CCD frequency to the frequency the objects create or reflect. Likewise, we need to defeat the camouflage or cloaking capabilities of the object to detect its presence. Hence, I use the sun. As stated earlier, I believe cloaking is possibly used by ETV occupants since we use it today for scientific displays of our own technology. However, I believe it is not possible to cloak the actions of energy transfer or radiation when powering a spacecraft or driving it near the sun. What solar satellite imagery does is display both of those conditions and sometimes simultaneously.

A photographer can move a subject in front of the camera ensuring the right distance, the right light is used, and intensity of the light is correct for the shot. Everything is controlled to maximize the result. When shooting images in a chaotic environment such as space, no one gets the luxury of object placement, lighting angles or speed of objects. You simply control the apparatus used for capturing the image. For this reason, I have scanned over 3 million images from mainly three different solar satellite systems to find that just right shot when the subject/object is interacting with the solar plasma of the chromosphere illuminating what was previously hidden. It's all chance, but thankfully the satellites produce so much imagery data it is possible to capture a Technosignature within a chaotic scene if you know what to look for and have the patience to look through a mountain of data.

Creating Categories

Being able to create categories or a classification system within a new science is challenging to say the least. In the mental health field, the Diagnostic and Statistical Manual of Mental Disorders (DSM) is one qualitative classification system that required years of development because it is a qualitative process (subjective). Mental illness has been acknowledged since the dawn of time or since man could insult another man about each other's political views. However, it took mankind until last century to classify and catalogue these observations into what we now have in the DSM. Likewise, we have claims and observations of alleged ETVs for centuries but have little substantial data to create a classification system around it. In qualitative research a grouping of like items is collected, and a theme is then gleaned from this group. To ensure validity of a theme, others with institutional knowledge of the subject in question will repeat this process to compare their conclusions as to thematic content. If enough subject matter experts agree on a central theme for a group of items, then the theme can become offered as a category or new classification. Qualitative analysis by nature is subjective and requires solid operational definitions, strict adherence to scientific methodologies outlined before research begins and bias reduction to the lowest levels possible. Unlike a mathematics problem, there are no hard, fast absolutes in qualitative analysis. However, if enough equivalent data of similar objects can be identified and agreed upon as significant, then the process of categorization can begin.

Early on in my review of NASA SDO imagery I kept finding small objects crossing in front of or near the sun but in only a single frame during a sequenced image review. These objects had a trail in the shape of a tear drop in 90% of their appearances. The other 10% of these objects without any trail yet seemed to be the same type of object (see images below).

AIA Detector Frequency

| 94 | 131 | 171 | 193 | 211 | 304 | 335 | 1600 | 1700 |

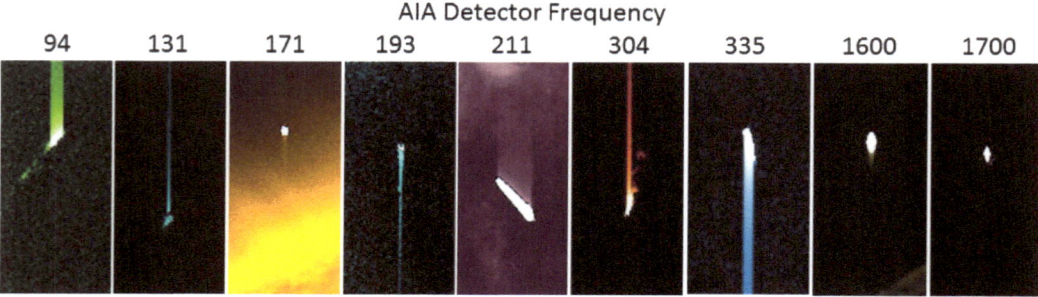

Fig. 26 Examples of Solar Centric objects from all the UV sensors on the SDO satellite.

I began adding tag descriptions to my date-time nomenclature of each picture for easy recall during a search of like items. My picture name would end up looking like this: 20180418_114354_4096_0304 <u>Solar centric</u>. The first series of numbers are the year/month/day group followed by the hours/minutes/seconds group. The next four numbers in this SDO file are the size of the image created indicating it is a 4K picture (4096). The last four digits list the angstrom or frequency used to create this image (0304). The term "Solar Centric" was my early attempt to assign a name for this new category of objects. It best described an object near the sun by proximity, but not by origin. In other words, these objects had nothing to do with the sun but are simply transiting the area at the time the shutter opened on a sensor. At the time of this writing, I have collected approximately 8,500 of these type objects in my personal image file. A quick note: when viewing these AIA sensor images, a predetermined color (like Red for 0304) is used to quickly identify the frequency band used by the CCD to create the image. Other colors are assigned to other angstroms but be aware the raw data before colorization occurs is black and white with 255 shades or intensity levels in between them for all ultra-violet frequencies captured by CCDs.

Within this new category, I called "Solar Centric" objects, I found a subcategory linked by shape. The unique features of this new subcategory were the object appeared in a single image, the object had a symmetrical shape, and the shape was a pattern of white dots and lines seemingly related and very organized. White as a color remember is the extreme end of intensity of any color regardless of color assignment (see figure 26 above – page 60). This means the sensor can accept no more energy into the pixel and reached its electrical maximum. The object is creating a frequency intensity beyond what the instrument can handle. The sun in each of the images in Figure 25 displays frequency intensities between minimum and maximum allowed by the sensor to show differences in surface texture. However, the figure below (page 62) display an area of maximum intensity (white) greater than the sun for that specific frequency. A collection of images of this new subcategory of solar centric objects are seen in Figure 27 below (page 62). Each object in the series of pics below was imaged on different days, times, and years. This new subcategory has yet to be named but is obvious the pattern displayed by all of them is similar.

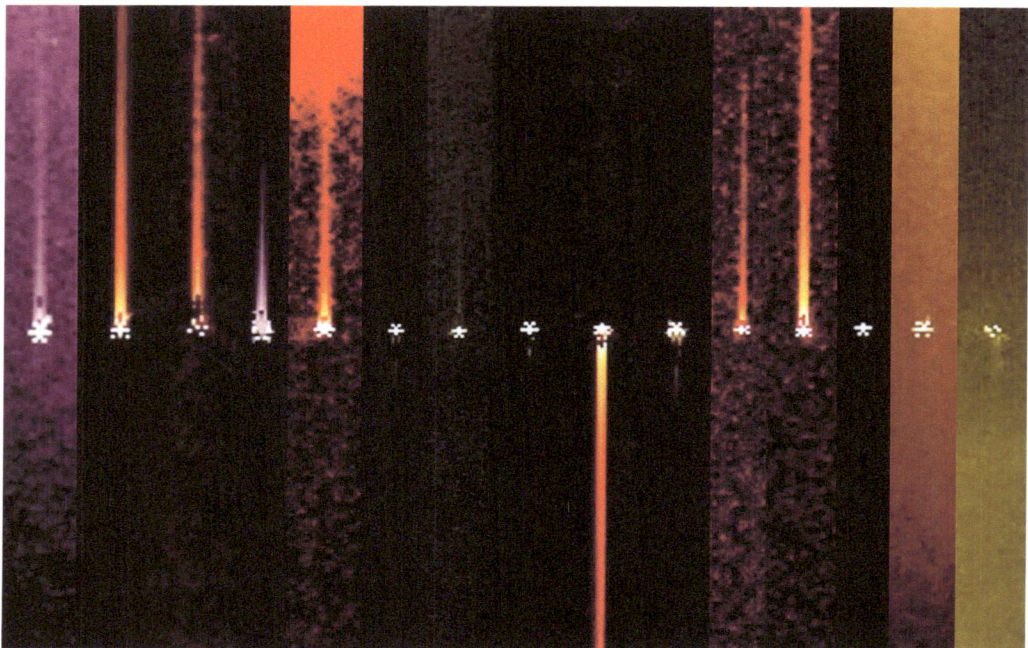

Fig. 27. Examples of a new subcategory of Solar Centric objects developed from the UV sensors on the SDO satellite. (Courtesy of NASA SDO satellite archives)

The objects in Figure 27 have the unique characteristic of symmetric shape and highly organized patterns. No other example of an asteroid or comet can produce this pattern and speed with which the object transverses the sensors field of view in a single exposure (regardless of cadence). Each of the 15 sample objects in Figure 27 came from a different time or date within the SDO archive. The different colors indicate different sensors looking at different frequencies were detecting the same kind of objects. This is just another example of finding subcategories from general categories that will continue as this exploration advances.

Mixed Methods Statistical Analysis

When attempting to prove a theory or determining a hypothesis is valid, it is best accomplished using a statistical model that holds a high degree of power or a high degree of validity. I spent four semesters in my first Master of Science degree in Human Factors at Embry-Riddle learning statistics. I have a new-found respect for this type of mathematical expression as it can make you seem like a genius for those who don't understand science or a fraud for those who do. It is equivalent to mathematical nitroglycerin. Statistics can be very unstable for those who are not highly skilled in what I consider the dark arts of math. However, when deciding to write my Graduate Capstone Project paper, I was forced to handle this type of math as a methodology to determine if my hypothesis would fall into the "Null" (expected) or "Alternative" side. In other words, could I prove my theory that the objects found, specifically in the COR2 STEREO – Ahead were intelligent by way of a statistical test.

The Mixed Method approach in statistical analysis means you start from a qualitative (subjective) method and move to a quantitative (objective) method. I gathered my pictures together into a separate file of all the COR2 images that contained an object (or light) sequence of 3, 6 or 9 segments as depicted in the image as seen in Figure 28 below (page 64).

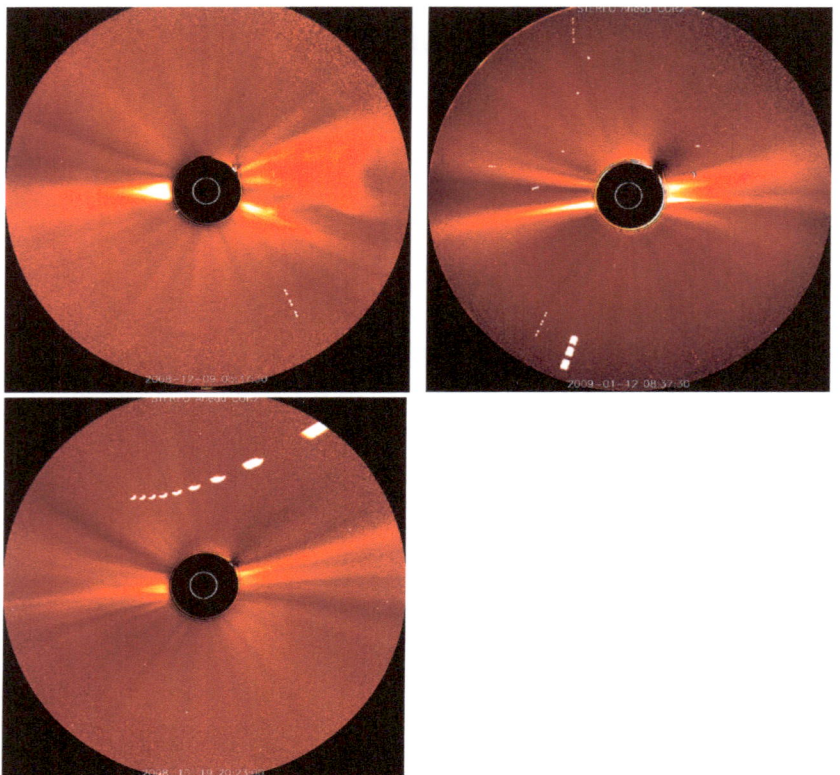

Fig. 28. Three STEREO – Ahead satellite images. (Courtesy NASA STEREO – Ahead Satellite archive)

I didn't need to know what the objects were but that these objects had the same optical characteristics comparatively. Out of a file of similar images I eventually collected 248 different pictures with seemingly similar objects transitioning through the image. The time span from my first image to my last in this file was 398 days. Because NASA was so meticulous in date-time stamping each image I was able to easily convert my file of qualitative data into a series of dates and time thus turning the data into a quantitative series of numbers. For example, each time a common unknown object was seen near the STEREO-A satellite, a date/time was assigned to the image and archived. Starting from image one (Day 0.0) each event was then listed. This list then became what is known as a bivariate string of data (having two columns).

# Days	Time
0.0	4:07:30
5.0	5:07:30
10.0	8:37:30
26.0	9:53:20
31.0	16:53:20
38.0	3:53:20
38.5	16:53:20
39.0	3:23:20
41.0	18:23:20
42.0	23:53:20

From this bivariate list one can create a graph called a Histogram. A Histogram is a graphical display of a series of equally divided sets of data (commonly a bar group) within a bin (entire graph). In a histogram, each bar groups numbers into ranges. The taller the bar, the more data was found in that time frame or frequency. Over the full set of time measured, sampled data is placed into the appropriate date/time groups or bars. These bars allow a visual indicator of how many items fall into a particular bar before a new bar or slice of time starts the process over again. Histograms end up looking like an amplitude modulated frequency where time is a constant, but object appearance counts can vary dictated by the archived image date/time stamp. If these COR2 objects always appeared twice every week, then if the bars on the graph equaled seven days each, the all the bars would indicate two objects in frequency with all the bars having the same height (see Table 1 – page 66)

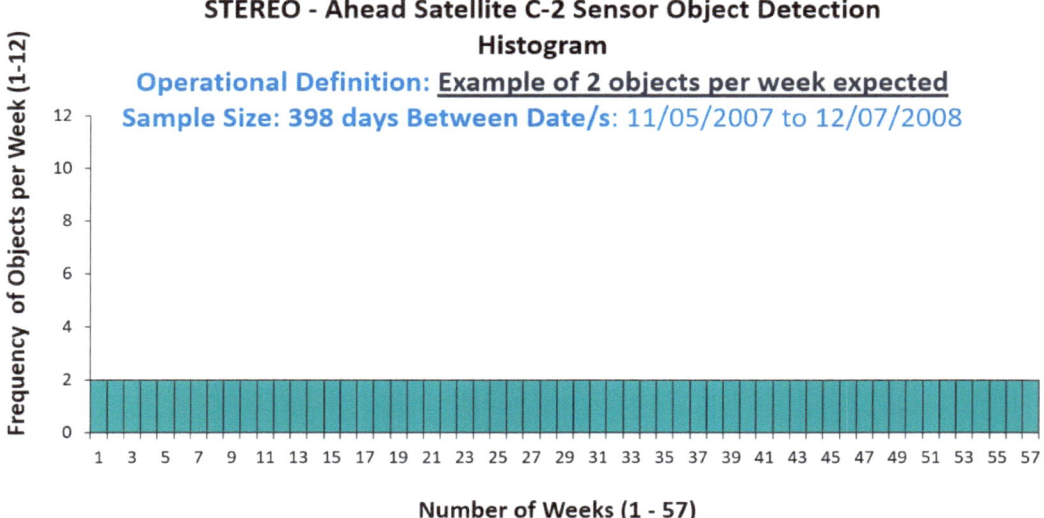

Table 1 - This histogram shows no variation of objects expected when assuming two objects will be the result of the investigation (hypothesis).

Once I had entered the dates and times into an Excel spreadsheet, I was able to document down to the second when each picture was produced as these objects appeared nearby. What is important about collecting pictures with time on them? An image is qualitative where time is quantitative. In the raw form a newly created qualitative data set does not usually contain any recognizable significance on its own. However, once the data is entered from the set being investigated, a histogram can offer a visual result of trends as a method of analysis. This is the histogram result using the COR-2 sensor from the STEREO – Ahead satellite (See Table 2 – page 67).

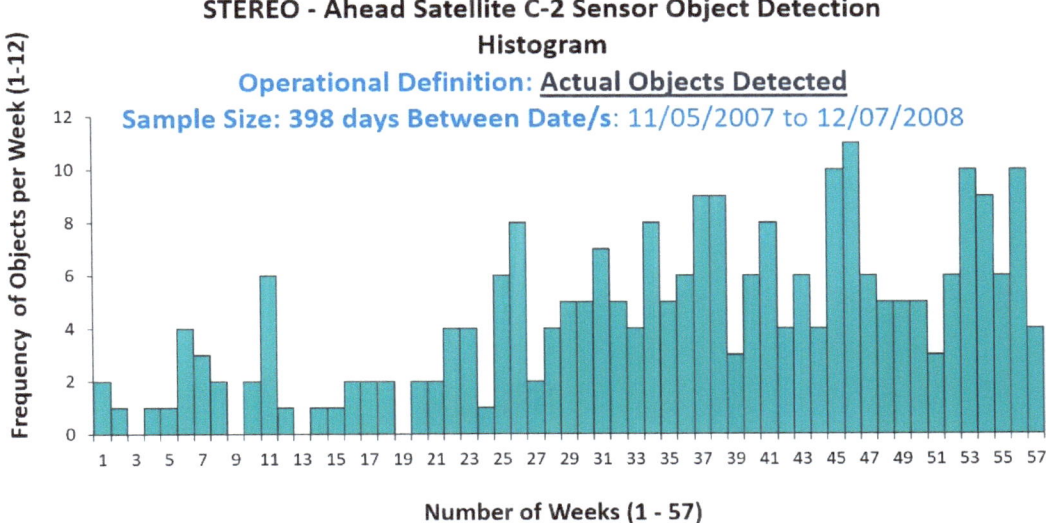

Table 2 - This histogram shows the variation of objects actually discovered within the data set of Technosigatures found.

This is where statistical analysis comes in and offers the researcher a way to make the data reveal deeper patterns of information if they exist! Remember, testing data to determine how a hypothesis is supported can go either for or against the original question or hypothesis. Using the tried-and-true scientific method of observation, measuring and testing can we obtain new information about this subject of Technosignature discovery.

To keep most readers from academic shock I will refrain from diving deep into the dark world of statistical analysis. However, the world of statistics can offer researchers answers they are looking for or a direction to go for deeper understanding. When handled correctly, statistics is a powerful tool that can determine the validity of a theory such as my hypothesis that Technosignatures are real and have at least one attribute of intelligence that can be assigned to them. In my case (writing my Graduate Capstone Project paper) this intelligent attribute turned out to be a partial sinusoidal wave. Further statistical analysis resulted in a pattern of predictable object arrival with a p-value of 7.7×10^{-27}. For context, a p-value of 5×10^{-2} is considered significant. The smaller the p-value, the more significant the result. My result was extremely significant, not proof, but close to it.

Object Attributes and Some Cool Pics

Once an object is determined to have enough attributes to be considered a Technosignature, the attributes themselves must present something unique as compared to common astronomical objects. For instance, comets and asteroids are usually present as a bright, spherical shaped object within the solar satellites' bank of sensors and often have a tail or plasma trail that follows. These objects are captured on a series of images due to their distance from the sensors and large field of view of the sensor itself. However, when an object is seen in only one frame among many in a series, the assumption is it is either close to the sensor or moving at extreme speeds far away.

A website called Solar Suspicions (www.solarsuspiocions.com) contains over 1,200 images with unusual attributes that have earned their way into the Technosignature category. Most of these images are found in only one in a series of images taken for that specific sensor. Some of the images on the web site show plasma trails fading behind the sun indicating the object is also further away than the sun from the sensor. Some objects show an interaction with the solar surface or atmosphere. These objects can be compared directly to the sun for size since the distance is now known. That said, the next few images are to be considered as true Technosignatures by their very nature of speed, shape, and size.

Fig. 29 Courtesy of NASA SOHO Image Archive (https://sohowww.nascom.nasa.gov/data/data.html). Image date/time/sensor: 19970201 0553 C2 (also found at www.solarsuspicions.com).

Figure 29 above (page 68) shows an object with a startling pattern of lights that are highly organized in an intelligent pattern. This object was captured in only one image near the edge of the field of view. Distance and size are unknown, but speed can be assumed as extreme as it was not found in subsequent images. The main reason this object is considered a Technosignature is the intelligently organized pattern of lights.

Fig. 30 Courtesy of NASA SOHO Image Archive (https://sohowww.nascom.nasa.gov/data/data.html). Image date/time/sensor: 20000505 1536 195

Figure 30 shows an extremely large object moving near the solar surface. This frame is one of three images where this object could be seen approaching the sun from the upper right moving to the lower center of frame. The object never touched the surface nor disturbed the chromosphere in a significant manner.

Fig. 31 Courtesy of NASA SDO Image Archive (Index of /assets/img/browse (nasa.gov). Image date/time/images size/sensor: 20130224_230140_4096_0335

Figure 31 shows an object in only one frame, with a trail of light coming from an object displaying a series of bright dots along an "L" shaped craft. The bright dots along the crafts' outer structure are not smeared as expected for a CTE condition (CCD anomaly/artifact) but leaving open the possibility the trail behind the craft is some sort of excited plasma response to the craft or exhaust from the craft.

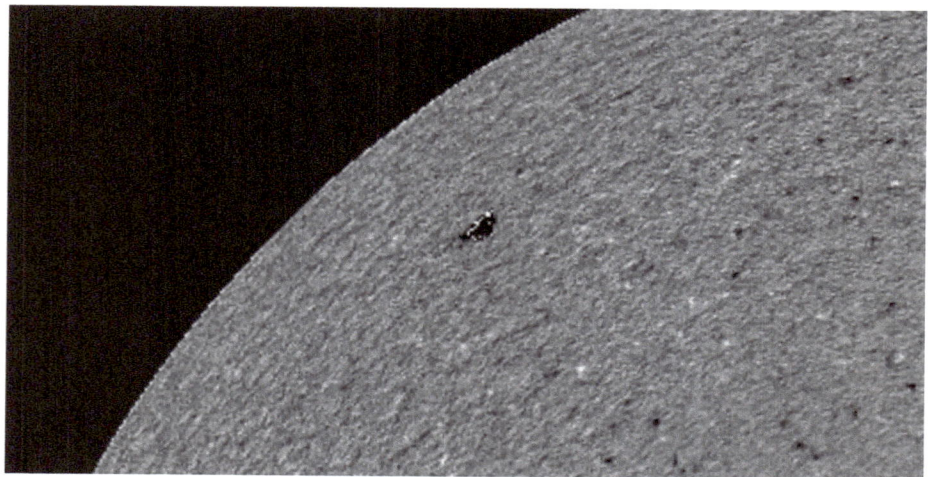

Fig. 32 Courtesy of NASA SOHO Image Archive (https://sohowww.nascom.nasa.gov/data/data.html). Image date/time/sensor: 20051225 1427 HDMI (also found at www.solarsuspicions.com).

Figure 32 (page 70) shows an object in only one frame, taken by the magnetic sensor where black indicates maximum magnetic north and white indicates maximum magnetic south (as traditionally understood). The suns' surface is shown as various shades of gray indicating a low-level blend of magnetic conditions. However, the triangular object between the sensor and the solar surface indicates extreme magnetic polar conditions outlining the shape of the craft and organized points of magnetic design. The distance and size of this craft is unknown but the extreme magnetic polarities of black and white intelligently organized and outlining the crafts' shape makes this object a good Technosignature candidate.

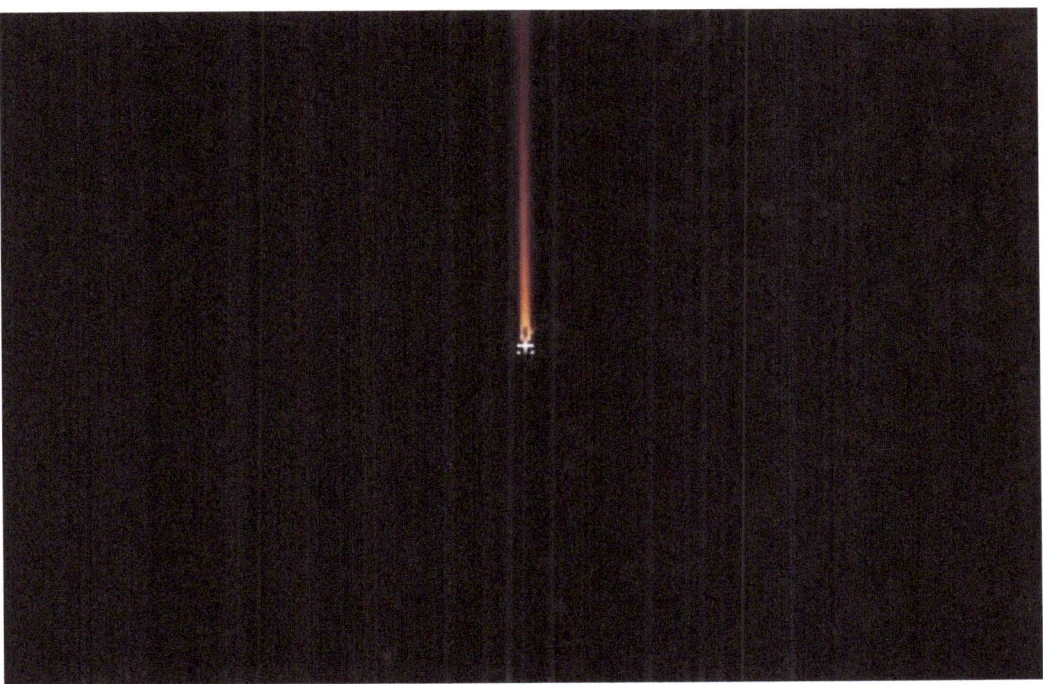

Fig. 33 Courtesy of NASA SDO Image Archive (Index of /assets/img/browse (nasa.gov). Image date/time/images size/sensor: 20130604_153020_4096_0304

Figure 33 shows an object in only one frame, with a trail of light streaming from a symmetrical object. The size and distance are unknown but the perfect symmetrical shape leaves little doubt this object was intelligently made. The figure below shows many similar objects from different dates and times and sensors.

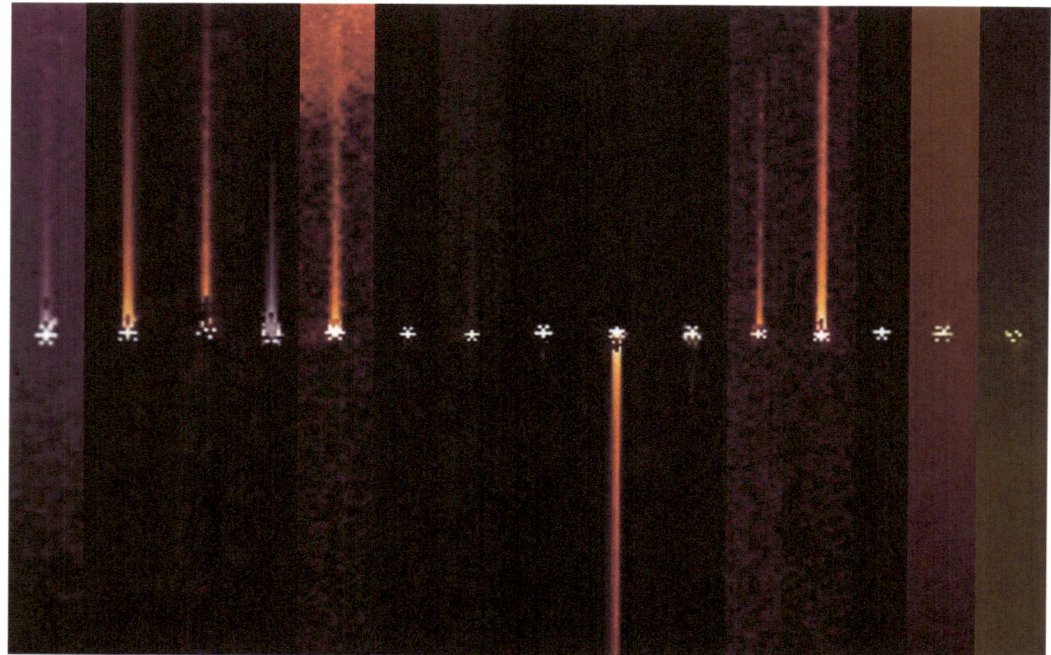

Fig. 34 Courtesy of NASA SDO Image Archive (Index of /assets/img/browse (nasa.gov).

Figure 34 shows multiple similar objects, some with a trail of light streaming from behind it and some without streams or trails. These objects were consolidated to show the slight differences in shapes and patterns but also show each one was intelligently designed. When a large collection of qualitative objects with similar attributes are brought together, then a quantitative analysis can be done using numerical count of objects within a specified timeframe.

Fig. 35 Courtesy of NASA SDO Image Archive (Index of /assets/img/browse (nasa.gov). Image date/time/images size/sensor: 20170403_144109_4096_211193171rg

Figure 35 displays a layered image using three sensor frequencies (211, 193, 171 angstroms) on top of each other. This is one of several images of a CME (Chronal Mass Ejection) where an object appears once passing through the CME in the upper right corner of the FOV. When zoomed in, the object appears to be solid and intelligently designed. Those who have commented on this object equate the shape of Star Trek's Bird of Prey (as seen below).

Fig. 36 Star Trek's bird of Prey (Courtesy of Pinterest.com)

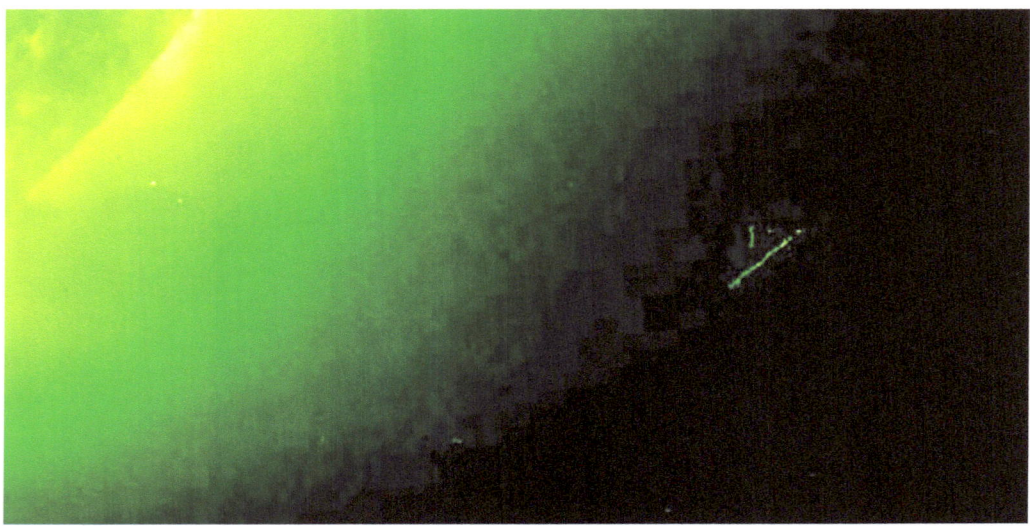

Fig. 37 Courtesy of NASA STEREO Ahead Image Archive (Search for STEREO Images NASA.gov). Image date/time/sensor: 20090602-150530-195

Figure 37 shows an unusually shaped object that appears in only one frame. The shape of the craft is what makes this one unique. The size and distance of this object are unknown. One of the Gestalt principles (Closure) is at play with this object is the partial edges outline what is assumed to be a solid structure in between the lines.

Fig. 38 Courtesy of NASA STEREO Ahead Image Archive (Search for STEREO Images NASA.gov). Image date/time/sensor: 20090504-122830-171

Figure 38 Displays a series of dots and lines of light in proximity to each other making this shape unknown and unique. Given the dark background the crafts' shape is also unknown but the intense white lights of the lines and dots indicate the sensor has reached its maximum ability to measure the incoming intensity for this particular frequency (This shows the object generating more frequency intensity than the sun is producing within the same FOV).

Fig. 39 Courtesy of NASA SDO Image Archive (Index of /assets/img/browse NASA.gov). Image date/time/images size/sensor: 20160727_011054_4096_0193

Figure 39 shows this object has a unique light-bar that seems to be associated with a nearby structure or as a single structure moving together as one. This craft is of unknown size and distance but does contain structure detail that is rare in this type of work. The vertical smear is not from CTE due to its short trail offering the possibility the light trail is some sort of exhaust or plasma excitation.

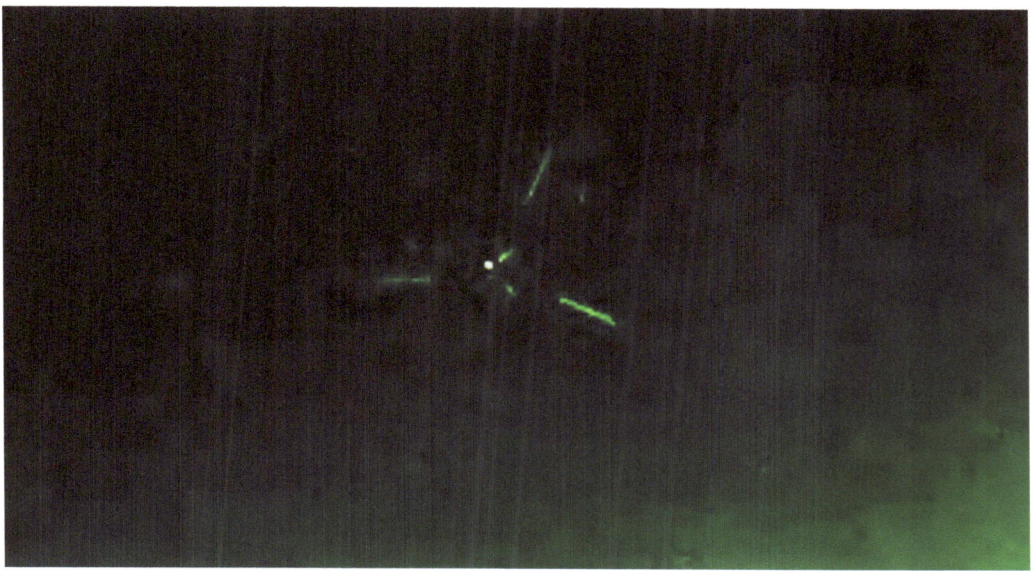

Fig. 40 Courtesy of NASA STEREO Ahead Image Archive (Search for STEREO Images (NASA.gov). Image date/time/sensor: 20091226-103030-195

Figure 40 shows a unique pattern found only in one frame containing dots and lines of light arranged in an intelligent pattern. The distance and size of this object is not known but the organized proximity of elements making up scene leaves little doubt this anomaly contains Technosignature attributes worth further investigation.

In Closing

Only a handful of educational institutions offer a four-year BS degree in Astrobiology such as Florida Institute of Technology and Arizona State University. Only one, Penn State, offers a dual Ph.D. program for Astrobiology if you list another traditional focus area such as Chemistry or Microbiology. Otherwise, the Astrobiology field is considered a very immature field of study. Even so, it is expected that you immerse yourself in astronomy, astrophysics, electronics, computer science and have research prowess at the Master of Science or Ph.D. level before claiming to be an Astrobiologist. The reason is, skipping over these fields of study will swallow your time in chasing simple mistakes of proper Technosignature identification and categorization. Knowing what right looks like within space imagery is extremely critical when comparing suspicious objects found through satellite sensor systems. Many strange anomalies are consistently found in satellite imagery, but one needs to have the experience and tools to determine which ones are common or natural and which ones need further investigation.

By staying alert when reviewing space imagery, it is possible the evidence of more Technosignatures is in front of you. Becoming familiar with typical satellite issues where camera/computer artifacts are unintentionally created and how they are made will keep one from making embarrassing claims. Handle science with care but stay away from dogma by religious or educational institutions ready to bias on what you are working. Remember, Astrobiology is a young field with many eager to shoot down anything controversial regardless of how much data you present to support your work. Just be sure your work is well vetted, accurate, and can be defended easily. A lot of study must be done before posting to YouTube your findings for the first several months if not years. If you find yourself staring the images at www.solarsuspicions.com, it is because I have spent 8 years studying and collecting objects with the right attributes for Technosignature consideration. I then created this website to encourage open discussion with enough evidence for both amateurs and professionals to use as a starting point for the next level of satellite imagery analysis and investigation. I hope enough information was presented for consideration that we are not alone in this universe. As astrobiology matures and more information is collected the trend pointing to life off-planet is possible with proof they are also being detected by Earth focused weather satellite like the Alaska focused satellite image below. For your consideration, please see figure 40 below.

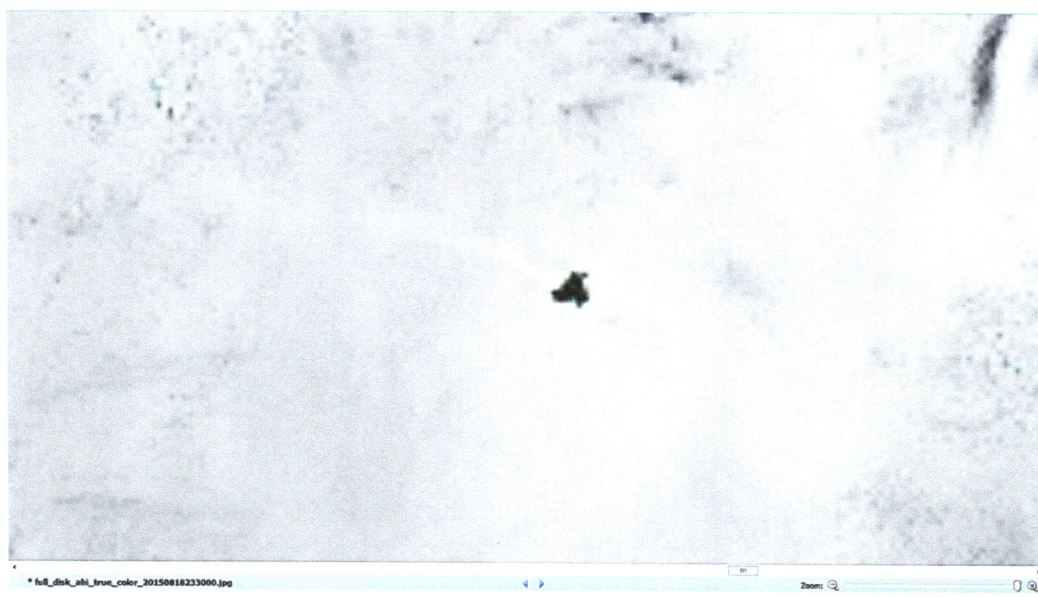

Fig. 41 Courtesy of Colorado University - Regional and Mesoscale Meteorology Branch (aa full_disk_ahi_true_color_20150818233000 015)

Figure 41 is an Alaska weather satellite image taken from a series of otherwise empty frames except this one containing an oddly shaped and colored object with a slight trail (heat or plasma) following the direction of travel. My military aviation background brings me to the conclusion this object is highly unique and possibly off-world given its shape and color. The objects' size and distance from the satellite is unknown but its shape is what leaves the observer with more questions. Welcome to astrobiology.

References:

Aschwanden, M. J. (2010). Image Processing Techniques and Feature Recognition in Solar Physics. Solar Physics, 262(2), 235–275. http://doi.org/10.1007/s11207-009-9474-y

Byrne, J. P., Morgan, H., Seaton, D. B., Bain, H. M., & Habbal, S. R. (2014). Bridging EUV and white-light observations to inspect the initiation phase of a "two-stage" solar eruptive event, (June), 1–22. http://doi.org/10.1007/s11207-014-0585-8

Banda, J. M., Angryk, R. A., & Martens, P. C. H. (2013). Steps Toward a Large-Scale Solar Image Data Analysis to Differentiate Solar Phenomena. Solar Physics, 288(1), 435–462. http://doi.org/10.1007/s11207-013-0304-x

Hurlburt, N., Cheung, M., Schrijver, C., Chang, L., Freeland, S., Green, S., … Timmons, R. (2010). Heliophysics Event Knowledgebase for the Solar Dynamics Observatory and Beyond. Solar Physics, 275(1-2), 17. http://doi.org/10.1007/s11207-010-9624-2

Ipatov, S. I., A'Hearn, M. F., & Klaasen, K. P. (2007). Automatic removal of cosmic ray signatures in Deep Impact images. *Advances in Space Research*, *40*(2), 160–172. http://doi.org/10.1016/j.asr.2007.04.012

Knut H. Stamnes, Wei Li, Hans Arthur Eide, Jakob J. Stamnes, "Challenges in atmospheric correction of satellite imagery," Opt. Eng. 44(4) 041003 (1 April 2005) https://doi.org/10.1117/1.1885469

Lillesand, Thomas M., Ralph W. Kiefer, Johnathan W. Chipman, (2015). *Remote sensing and image interpretation.* Seventh Edition Wiley, Ch. 5 (283-382), Ch. 7 (485-602) ISBN-13: 978-1118343289

Lindler, D. J., A'Hearn, M. F., Besse, S., Carcich, B., Hermalyn, B., & Klaasen, K. P. (2013). Interpretation of results of deconvolved images from the Deep Impact spacecraft High Resolution Instrument. Icarus, 222(2), 571–579. http://doi.org/10.1016/j.icarus.2012.09.003

Lindsay Hays et al. 2015. The NASA astrobiology roadmap. Astrobiology 15(4): 715-30, DOI:10.1089/ast.2015.0819

Mackay, D. H., Karpen, J. T., Ballester, J. L., Schmieder, B., & Aulanier, G. (2010). Physics of solar prominences: II - Magnetic structure and dynamics. Space

Science Reviews, 151(4), 333–399. http://doi.org/10.1007/s11214-010-9628-0

Martens, P. C. H., Attrill, G. D. R., Davey, a. R., Engell, a., Farid, S., Grigis, P. C., … Timmons, R. P. (2012). Computer Vision for the Solar Dynamics Observatory (SDO). Solar Physics, 275(1-2), 79–113. http://doi.org/10.1007/s11207-010-9697-y

Pantazis, D., Oliva, A., & Cichy, R. M. (2014). Resolving human object recognition in space and time. Nature Neuroscience. http://doi.org/10.1038/nn.3635

Saad Saoud, T., Moindjie, S., Autran, J. L., Munteanu, D., Wrobel, F., Saigne, F., … Glorieux, M. (2014). Use of CCD to Detect Terrestrial Cosmic Rays at Ground Level: Altitude vs. Underground Experiments, Modeling and Numerical Monte Carlo Simulation. *IEEE Transactions on Nuclear Science*, *61*(6), 3380–3388. http://doi.org/10.1109/TNS.2014.2365038

Smith, A. R., Mcdonald, R. J., Hurley, D. L., Holland, S. E., Groom, D. E., Berkeley, L., Wei, M. (2002). Radiation events in astronomical CCD images *, *4669*, 172–183.

Snyder, D.L., Hammond, A.M., White, R.L., 1993. Image recovery from data acquired
with a charge-coupled-device camera. J. Opt. Soc. Am. A 10, 1014–1023.

Turmon, M., Jones, H., & Malanushenko, O. (2010). Statistical Feature Recognition for Multidimensional Solar Imagery. Solar Physics, (May), 1–24. http://doi.org/10.1007/s11207-009-9490-y

Turmon, M., Pap, J., & Mukhtar, S. (2002). Statistical pattern recognition for labeling
solar active regions: application to SOHO/MDI imagery. The Astrophysical Journal, 396–407. Retrieved from http://iopscience.iop.org/0004-637X/568/1/396

Van Coillie, F. M. B., Gardin, S., Anseel, F., Duyck, W., Verbeke, L. P. C., & De Wulf, R. R. (2014). Variability of operator performance in remote-sensing image interpretation: the importance of human and external factors. International Journal of Remote Sensing, 35(2), 754–778. http://doi.org/10.1080/01431161.2013.873152

Van Dokkum, P. G. (2001). Cosmic-Ray Rejection by Laplacian Edge Detection. *Publications of the Astronomical Society of the Pacific, 113*(789), 1420–1427. http://doi.org/10.1086/323894